FUNDAMENTALS OF ADDITIVE MANUFACTURING FOR THE PRACTITIONER

FUNDAMENTALS OF ADDITIVE MANUFACTURING FOR THE PRACTITIONER

FIRST EDITION

Additive Manufacturing Skills in Practice Series

SHEKU KAMARA

Milwaukee School of Engineering

KATHY S. FAGGIANI, PHD

WILEY

Registered Office
John Wiley & Sons, Inc., 111 River Street, Hoboken, NJ 07030, USA

Editorial Office
111 River Street, Hoboken, NJ 07030, USA

For details of our global editorial offices, customer services, and more information about Wiley products visit us at www.wiley.com.

Wiley also publishes its books in a variety of electronic formats and by print-on-demand. Some content that appears in standard print versions of this book may not be available in other formats.

Library of Congress Cataloging-in-Publication Data is Available:

ISBN 9781119750383 (hardback)
ISBN 9781119750505 (ePDF)
ISBN 9781119750512 (ePub)

Cover Design: Wiley
Cover Image: © Sandra M/iStock/Getty Images, kynny/iStock/Getty Images, lucadp/iStock/Getty Images

Set in 9/13 pt STIXTwoText by SPi-Global, Chennai, India

SKY10030323_100421

Contents

Chapter 1

Introduction: Moving into Additive Manufacturing

Case Introduction: Current work roles in manufacturing and how they change in AM

Great West Manufacturing (GWM), a medium-size manufacturing firm, recently experienced a change in leadership. The board of directors has charged the new CEO, Sherman Potter, to explore and implement additive manufacturing (AM) to better prepare the firm for long-term viability in their increasingly competitive manufacturing space. The new CEO has made it clear that GWM will continue to manufacture the current line of recreation and athletics end-user products but would like to expand operations to include new opportunities facilitated by AM. As part of the exploration effort, Potter assigned Pete Granger (manufacturing process engineer), Bob Nelson (design and materials engineer), Edgar Remmins (manufacturing technician), and Roxanne Jensen (compliance, testing, and quality control engineer) to the AM Pilot Project team. Their charge is to investigate the knowledge and skills

Fundamentals of Additive Manfacturing for the Practitioner, First Edition. Sheku Kamara and Kathy S. Faggiani © 2021 John Wiley & Sons, Inc. Published 2021 by John Wiley & Sons, Inc.

needed to support additive manufacturing at GWM and formulate a plan to conduct an additive manufacturing pilot test at GWM.

In their first meeting, each team member revealed that s/he had minimal understanding of additive manufacturing with some familiarity with 3D printing concepts. The team's experience in traditional manufacturing ranged from 2 to 24 years in their respective career areas. The group decided to start by preparing a list of GWM's manufacturing roles and responsibilities to compare current talent to the knowledge and skills needed for additive manufacturing. Their focus would then shift to exploring additive manufacturing and 3D printing (AM) requirements and processes in preparation for developing a proposal to initiate a pilot project within GWM.

INTRODUCTION

Imagine the year is 2025. The world has quelled COVID-19, the United States has convinced many American manufacturers to move production back to the states, and focused financial investments have spurred significant ongoing growth in the manufacturing sector. Simultaneously, the manufacturing workforce has suffered a hit from the loss of seasoned employees through retirement. Manufacturing has had limited success in recruiting high school and college graduates and is struggling to find employees with the right skillsets to advance additive manufacturing and other emerging production approaches. Fortunately, many great opportunities for exciting work and career growth exist for those who choose a manufacturing career path. While you may have significant experience in different aspects of manufacturing as a manufacturing practitioner, are you ready to meet the challenge of implementing additive manufacturing technologies and processes in the next four to five years?

In the early 1980s, Chuck Hall produced the first stereolithography part, and the following year filed a patent for the Stereolithography Apparatus (SLA). Figure 1.1 displays the first SLA machine and printed part. Additive manufacturing, previously termed rapid prototyping, was officially started when 3D Systems introduced the first commercially available AM system, the SLA-1, in 1987. Various terms referred to additive manufacturing until 2009 when the ASTM F42 committee formed supported by SME's RTAM community. In 1993, researchers at the Massachusetts Institute of Technology (MIT) patented "three-dimensional printing techniques," which specifically referenced binder jetting (Sachs, Haggerty, Cima and Williams, 1993). The patent defined the process as:

> . . .making a component by depositing a first layer of a fluent porous material, such as a powder, in a confined region and then depositing a binder material to selected regions of the layer of powder material to produce a layer of bonded powder material at the selected regions. (US Patent US5204055A, Abstract)

Three-dimensional printing, abbreviated as "3DP" or "3D printing," has evolved and is now used to refer to inkjet-based, low-cost, hobbyist 3D printers. 3D printing is considered synonymous with additive manufacturing. In general, industry insiders use additive

Figure 1.1 First Stereolithography Machine and First 3D Printed Part
Source: Courtesy of 3D Systems, Inc.

manufacturing or AM, and the public typically uses 3D printing or 3DP to refer to the industry. For this book, both terms refer to the broad range of technologies and processes that comprise additive manufacturing.

Additive manufacturing and 3D printing (AM) technologies began to appear in the 1980s and have since become a significant force for revolutionizing product design, production process efficiency and effectiveness, supply chains, and innovation processes. According to Grand View Research (2020), the global AM market will exceed $35 billion by 2027. North America will likely maintain the largest market share of around 35%.

The primary business drivers for rapid growth and investment in AM continue to be enhanced product manufacturing and reduced time to market.

Continued growth and proliferation of AM are expected, with two key challenges that may hamper growth over the coming decade:

1. An overzealous focus on prototyping and low-volume production prevents organizations from realizing AM benefits at all phases of the manufacturing process.

2. Lack of a skilled workforce hampers the opportunities enabled by evolving technologies and processes.

Misguided Focus Prevents Realization of AM Benefits

AM was initially used in prototyping and to produce complex customized or low-volume production parts. These focused applications of AM led to the limited adoption of additive manufacturing technologies. For example, consider traditional injection molding, where a liquid polymer injects into a mold. When evaluated for use in injection molding, the AM focus tends to address increased part complexity, eliminate the mold, or incorporate conformal cooling, multimaterial or gradient-material usage, and speed. There is frequently a failure to consider the application and benefits in the other injection molding elements enabled by AM. In many circumstances, additive manufacturing is critical in providing elaborate fixtures, conformal cooling, mold inserts, and other support areas of the injection molding process. The aerospace and medical industries, which focused on prototyping in their initial AM application, provide another example. These industries realized significant cost savings and reduced lead-times by leveraging the technologies for design verification or presurgical models. However, other significant advantages exist for organizations that push past this initial narrow scope of AM applications.

The shift to AM requires considerable capital investment. For manufacturing firms to adopt these evolving technologies, key decision makers need to understand and leverage the broad range of benefits from more widespread use and application of AM technologies across the entire manufacturing process. Industries will only realize the value-add of AM through the work of a well-prepared and innovative workforce who can make it happen!

Lack of Skilled Workforce Limits Ability to Take Advantage of Opportunities

A skilled workforce is an essential component to exploit AM in any manufacturing and production setting fully. It is the existence of qualified employees who understand how to apply these technologies to current processes to improve and innovate. Research in over 400 manufacturing firms by Deloitte and The Manufacturing Institute revealed that the number of new manufacturing jobs would grow by almost 2 million workers by 2028 (Giffi et al. 2018). Furthermore, more than half of open positions in 2028 could go unfilled due to boomer retirements, misperceptions of manufacturing work, and new skillsets required to work with advanced manufacturing technologies.

In response to the well-publicized manufacturing skills gap and workforce shortages, various education and training initiatives in AM have developed. However, most of the efforts have attracted new generations of workers through high school, two-year trade and technical school, and four-year university programs. Current manufacturing practitioners rely on vendor-specific training, professional association workshops, on-the-job experimentation and training, and certification opportunities as venues for acquiring the necessary AM skills. Unfortunately, the efforts intended to address the manufacturing skills gap appear to have fallen short. A recent update to the Deloitte skills gap study indicated that workers' shortfall for available manufacturing jobs might be even higher than the original 2 million estimated (Wellener et al. 2020). A national desire to return the manufacturing of critical goods to the US following the COVID-19 pandemic may further exacerbate the manufacturing workforce shortage in the coming years.

This book prepares current manufacturing practitioners to move from traditional manufacturing practices to incorporating AM technologies and processes in their skillsets by providing a broad foundation for further learning. Existing additive manufacturing books and much available training for practitioners focus on the technical aspects of additive manufacturing and 3D printing technologies and techniques without assisting the practitioner in bridging the knowledge gap. This book's focus is to provide an end-to-end introduction covering all AM technologies and processes to allow the reader to develop a foundation for implementation in any manufacturing role while also broadening understanding of the scope of applications possible. Building a basic understanding of AM technologies and processes while providing broad coverage of potential applications will allow practitioners to more easily transition to the much-in-demand additive manufacturing workforce and leverage evolving technologies within their organizations.

Vital elements of additive manufacturing are introduced in upcoming chapters with enough technical details to provide the practitioner with a background to participate in the discussion of initiatives in their organizations and analyze the solutions presented. The remainder of this chapter covers an overview of manufacturing processes and examples of the unique applications facilitated by AM. A discussion of how traditional manufacturing job roles change in AM and suggestions for additional training and development follows. The end of the chapter provides a roadmap and recommendations for using the book.

AM Notable 1.1 - Are You Ready for Industry 4.0?

Industry 4.0, known as the new digital industrial revolution, integrates cloud computing, cognitive computing, the Internet of Things, and cyber-physical systems to automate and exchange data in manufacturing. This new industrial revolution enables unprecedented communication between products, means of production, and the humans involved. Software focused on traditional manufacturing limited AM's ability to reach its potential, but recent developments address AM's unique issues. HP and Dyndrite

recently announced an advanced software solution to support end-to-end AM processes and management solutions. It also connects to popular third-party software tools used in manufacturing (HP's Universal Build Manager Powered by Dyndrite, 2020).

What does that mean for the manufacturing workforce? It means almost every manufacturing employee will need to understand and use advanced systems in their work, so better sharpen up those technical skills!

MANUFACTURING PROCESSES

Manufacturing processes comprise four broad process categories: formative, subtractive, additive, and hybrid, used to produce a part or finished product. Figure 1.2 illustrates these categories. A general description of each type appears below, and specific examples of manufacturing processes included in each category are listed and described in Table 1.1:

| Formative | Subtractive | Additive | Hybrid |

Figure 1.2 Types of Manufacturing Processes

Source: Shutterstock.com (Formative, Subtractive, Additive); Hybrid image courtesy of Concurrent Technologies Corporation.

Table 1.1 Manufacturing Processes

Formative

Casting is the process of pouring liquid material, typically heated from solids, into a mold cavity to create the desired shape. This process is mostly used for metals. Investment and sand-casting processes are commonly used for metals.

Forming is the process of creating a desired metal part ramming the metal into the desired shape using a hammer or die. The material can be heated or at room temperature, depending on the process.

Molding is the process of pouring liquid material, heated from solids, into a mold under pressure, to create the desired shape. Die casting is for metals and injection molding for plastics. Both processes inject the material under high pressure and produces parts with superior surface finishes.

Stamping is the process of using a die, with compressive forces, to create a desired shape from a sheet metal. The process uses different forming techniques to produce complex parts.

Subtractive

Computer Numerical Control (CNC) is the process of using computer programming to control a cutting tool in precise movements to achieve a desired shape.

Drilling is the process of using a rotary cutting tool to create a hole into an object. The cutting tool referred to a drill bit, spins at different speeds, depending on the material, to remove material by shearing and extrusion. The material can be made of metal, plastic, or wood.

Milling is the process of removing material from an object using a rotary cutting tool to the desired shape. The object can be cut from different angles and axes.

Turning is the process of removing material from an object as it rotates to create a cylindrical part. The cutting tool is pressed against the rotating object to remove material vis shearing.

Table 1.1 *(Continued)*

Additive

Binder jetting is the process of fusing powdered material with a binder using print-heads. The material can be ceramic, plastic, or metal.

Directed energy deposition is the process fusing metal filament or powder using a laser or electron beam as the powder is sprayed or the filament extruded. The process can fuse multiple powdered material at the same time or in sequence using multiple nozzles.

Material extrusion is the process fusing metal or plastic material by heating and extruding the filament. Bonding occurs as the material cools down to a solid.

Material jetting is the process fusing liquid material, typically thermosets, by spraying droplets of the material through a printhead.

Powder bed fusion is the process of fusing metal or plastic powder using a laser or electron beam on a flat plate.

Sheet lamination is the process of fusing layers of material and using a cutter or mill to achieve the desired shape.

Vat photopolymerization is the process of fusing liquid material using an ultraviolet light, in form of a laser of projection, to create the desired from a vat.

Hybrid

Directed energy deposition + CNC is the process of fusing powdered metal using a laser within a CNC machine to build layers of the part whilst machining after a few layers to achieve the accuracy and surface finish of the subtractive process and higher complexity of the additive process.

Powder bed fusion + Milling is process of fusing powdered metal using a laser on a bed of powder and machining the inside and outside surfaces every 10 layers. This process allows the creation of conformal cooling channels within a mold and deep cuts within the part, only possible with other machining processes like electrical discharging machine (EDM).

Sheet lamination + CNC Milling is process of fusing thin metal foils using ultrasonic welding and machining the inside and outside surfaces every few layers. This process allows the creation of gradient material, including different metal combinations, internal cooling channels and capability to embed sensors and electronics within a part.

Source: Original, Kamara and Faggiani, 2020.

- *Formative.* Formative manufacturing processes include those that shape the material into the desired form and may use a mold or a die. Materials used may be melted, heated, or at room temperature, depending on the specific process. In general, formative techniques are used for high volumes of the same part and support lower part costs.

- *Subtractive.* Subtractive manufacturing processes remove material from a solid block or a near-net shape to achieve the desired size and geometry of the finished object. These techniques may use computer numerical control (CNC) machines, plasma torches, sheers, or other material-removal equipment. Parts or products produced by subtractive processes typically have simple geometry and are made in medium volumes.

- *Additive.* Additive manufacturing processes use computer-aided design (CAD) files to create ultra-thin digital slices of the desired shape, built-up by bonding layers of material together based on the slices to form an object. Additive manufacturing enables complex part geometry and typically produces parts in low to medium volumes.

- *Hybrid* Hybrid manufacturing processes achieve the desired shape of an object by combining either formative or subtractive processes with additive techniques sequentially within the same machine. This approach combines the complexity possible with additive methods and the superior finishing possible through other processes.

TRADITIONAL AND AM JOB ROLES

Manufacturing jobs generally include work in any process necessary to turn raw materials or components into new products. Broadly defined, classic manufacturing work falls into the following categories:

- *Process.* Work to select and implement the desired production process for parts or products; performed by a manufacturing engineer.

- *Design.* Work to conceptualize and design parts or products and optimize the design to leverage the selected process; performed by design engineers.

- *Material.* Work to identify and select the correct material for parts or products, particularly where the material must meet certain conditions; performed by material engineers.

- *Compliance.* Work to understand and ensure that parts or products meet ethical, safety, or regulatory standards; performed by a compliance engineer or compliance manager.

- *Testing.* Work to create a process or method to test and validate the part or product to ensure specifications are met; frequently performed by a test engineer or test analyst.

Various job titles exist within these general categories, and some roles may overlap among the types. Table 1.2 includes typical manufacturing job titles and brief descriptions from the US Bureau of Labor Statistics in the first two columns. The first column also shows new job titles for AM-specific roles emerging in job postings. Let's consider how traditional roles may need to evolve or be enhanced in additive manufacturing.

Table 1.2 How Additive Manufacturing Changes Job Roles

Role	Traditional Manufacturing	In Additive Manufacturing
Manufacturing Production Technicians **New Job Titles:** • Field Service Technician (AM)	Set up, test, and adjust manufacturing equipment.	• New technologies • New techniques (machine calibration, build preparation, part finishing, etc.) • New IT data and software skills
Manufacturing Engineers **New Job Titles:** • Additive Manufacturing Process Engineer	Design, integrate, or improve manufacturing systems or related processes.	• New technologies • More collaboration with commercial/design engineers • Rethink fabrication and modeling • Utilize DFAM (design for additive manufacturing) • New materials
Manufacturing Engineer Technologists **New Job Titles:** • Additive Manufacturing Technologist	Develop tools, implement designs, or integrate machinery, equipment, or computer technologies to ensure effective manufacturing processes.	• New technologies • Innovate with jigs/fixtures • More collaboration with manufacturing, manufacturing production technicians, and commercial/design engineers
Industrial Production Managers **New Job Titles:** • Senior Additive Manufacturing Engineer • Application Development Technical Service Manager (AM)	Plan, direct, or coordinate the work activities and resources necessary for manufacturing products in accordance with cost, quality, and quantity specifications.	• New technologies • Technical and management cross-functional competencies

(Continued)

Table 1.2 *(Continued)*

Role	Traditional Manufacturing	In Additive Manufacturing
Commercial and Industrial Design Engineers **New Job Titles:** • Senior Additive Manufacturing Engineer	Develop and design manufactured products, such as cars, home appliances, and children's toys. Combine artistic talent with research on product use, marketing, and materials to create the most functional and appealing product design.	• New design processes and technologies • New design programs • New materials • New IT data and software skills • Rethink fabrication and modeling • More collaboration with manufacturing engineers and production technicians • Utilize DFAM (design for additive manufacturing)
Finisher/Manufacturing Specialist **New Job Title:** Machine Operator	Use customer and design specifications to make products visually and dimensionally acceptable to consumer.	• New AM technologies or machines to enhance visual and dimensional properties, that both produce and finish parts • Perform both AM operator and finishing functions in same area
Machine Operators and Tenders **New Job Titles:** • Additive Manufacturing Technician	Set up, operate, or tend machines, such as glass forming machines, plodder. machines, and tuber machines, to shape and form products, such as glassware, food, rubber, soap, brick, tile, clay, wax, tobacco, or cosmetics.	• New technologies • New techniques (machine calibration, build preparation, part finishing, etc.) • New IT data and software skills • 3D scanning and non-destructive testing technologies • Part finishing, may replace separate finishing role
Material Engineers **New Job Titles:** • Materials Fabrication and Process Engineer • Materials and Process Engineer	Evaluate materials and develop machinery and processes to manufacture materials for use in products that must meet specialized design and performance specifications. Develop new uses for known materials.	• New materials • New parts/material/process certifications • New technologies

Table 1.2 *(Continued)*

Role	Traditional Manufacturing	In Additive Manufacturing
Quality Control Managers/Analysts **New Job Titles:** • Quality/Validation Engineer (AM)	Plan, direct, or coordinate quality assurance programs. Formulate quality control policies and control quality of laboratory and production efforts. Conduct tests to determine quality of raw materials, bulk intermediate and finished products.	• New IT data and software skills • New materials and processes • In areas where part consolidation is used, 3D scanning and other non-destructive metrology methods are a requirement to validate the parts.
Compliance Manager **New Job Titles:** • Senior VP, Quality, Compliance, and Regulation (AM)	Plan, direct, or coordinate activities of an organization to ensure compliance with ethical or regulatory standards.	• New parts/process/material certification
Tester/Inspector **New Job Titles:** • Additive Manufacturing Specialist	Inspect, test, sort, sample, or weigh nonagricultural raw materials or processed, machined, fabricated, or assembled parts or products for defects, wear, and deviations from specifications.	• New techniques and methods for more complex parts • In areas where part consolidation is used, 3D scanning and other non-destructive metrology methods are a requirement to validate the parts.

Source: Content for traditional job titles and work descriptions derived from O*Net OnLine Occupations. www.onetonline.org/ (accessed 11 May 2020).

The rise of additive manufacturing represents a paradigm shift in manufacturing and production. While the technologies and processes are groundbreaking and unique, the standard work categories remain the same. However, there are new knowledge, skill, and behavior requirements needed within each category. The new requirements are presented and described below, as well as summarized in Table 1.2.

New Technologies, Processes, and Materials

AM is comprised of a wide variety of technologies and processes that are related to different material capabilities, as described in the section on Manufacturing Processes. Given the proprietary nature of AM equipment and associated processes, each specific platform's education and training are vendor-provided. Specific process requirements may vary by equipment, material, or AM application to a given solution, and manufacturing engineers may need to think out of the box about modeling and fabrication. AM part finishing is automated or manual, with more significant automated finishing occurring to avoid the variation possible in manual processes. Material acquisition, storage and handling, safety issues, reuse or recycling, and other regulations vary by material and may differ significantly from traditional raw materials.

New Techniques

In addition to becoming experts with new machines and equipment, technicians and operators will need to apply modern machine calibration techniques and adjustments. New tasks and processes in preparing for the build of parts or products will be necessary and new methods for removing parts or products from AM machines and conducting various post-processing and finishing tasks.

New IT Data and Software Skills

Acquiring, recording, updating, and maintaining 3D data as input to the AM manufacturing process will present the opportunity to learn and apply new IT and data management skills. The ability to design parts or products with greater complexity for increasingly sophisticated systems will enable more nondestructive testing, quality control inspections with 3D scanning, and the generation of more useful data.

More Communication and Collaboration

AM can significantly reduce the time from design to production, resulting in a greater dependency between the design and engineering processes and technicians. The compressed time frame and dependencies indicate the need for more and better communication between design engineers, manufacturing engineers, and production technicians than may exist in some traditional manufacturing processes. Increased communication also minimizes errors due to unintended consequences. The speed of the process and greater complexity of designs account for some mistakes as operators may not distinguish between design intent and an error.

New Design Processes and Technologies

Additive manufacturing opens a new range of possibilities that make greater complexity in design and production easier to achieve; however, current traditional design thinking may create "blinders" for design engineers. Engineers will need to develop skills using new design processes and technologies to leverage AM process power.

The inherent variability within the AM process provides the opportunity for gradient and predictable mechanical properties instead of discrete, as is evident in other methods. For example, a change in laser power within the same part and on the same build platform produces composite properties using the same material as the parameters change layer-wise. The manufacturing engineer now must collaborate with design engineers, as part complexity and consolidation require the development of new processes and techniques. For example, six individual components consolidated into one part with increased functionality affects other processes downstream.

New Tooling Capabilities

The development of jigs and fixtures for additive manufacturing and more traditional processes is possible using AM. Manufacturing engineers and technicians will need to acquire the skills and knowledge to design and use AM-produced tools and fixtures rapidly. The increased level of complexity possible within AM has enabled designs for jigs and fixtures impossible through traditional methods.

Enhanced Use of 3D Scanning and Nondestructive Testing

A standard tool in modeling and design is poised to assume an even more significant role with AM's ability to produce increasingly complex objects. The 3D models created from scanning and modified for AM can facilitate reverse engineering of existing traditionally made parts. The complexity enabled by AM can reduce the number of individual components required in a part design. Design engineers will need to enhance their skills working with scanned images and reverse engineering. Quality verification of AM parts is best achieved by nondestructive testing (NDT), given the potential variability inherent in AM processes. Test and compliance engineers will need to become versed in NDT.

New Certification Requirements

Existing parts and products have been subject to quality, safety, and regulatory certification through production with traditional manufacturing products and materials. AM adds a whole new dimension, requiring new certifications for AM processes, materials, and parts. Compliance engineers, design engineers, and industrial production managers will need to work through required certifications.

AM Notable 1.2 - Can You Hear Me Now?

The disruptive change embodied in a move to AM doesn't involve only technology, process, materials, and the way work occurs. It also means that people and communication are more critical than ever. Advanced software can facilitate communication between equipment and processes, and even between technology and people, but the person-to-person connection is more important than ever. Sound workflow systems can alert workers

(Continued)

of pending jobs, smooth the production flow, and indicate anomalies before problems occur. But only strong human communication can help solve complex design problems, turn opportunities into creative innovations, and build a healthy work environment.

What does this mean for you? Don't focus so much on learning new AM technologies and processes that you ignore opportunities to develop your interpersonal communication and team skills. Soft skills matter – arguably more in an automated AM world than in traditional manufacturing, especially during the turmoil and ambiguity that accompany disruptive change.

In addition to the job roles and titles described in Table 1.2, new career opportunities are also emerging. For example, the job website SimplyHired.com listed an open position for an *Associate Data Engineer – Additive Manufacturing* (retrieved 12 June, 2020). According to the job description, the work involves using IT to collect, track, visualize, and access modeling, process metallurgy, and testing data for AM and materials science researchers. The second example of AM's emerging jobs is the *Additive Manufacturing Design Engineer,* a role involved with driving Design for Additive Manufacturing (DfAM) and AM-related solution development and exploration. An excerpt from the job description found at indeed.com (retrieved 5 November 2020) for an AM manufacturing design engineer appears in Figure 1.3. It is likely that as AM applications continue to expand and increase across industries, many new AM career opportunities like this one develop.

AM Notable 1.3 - Your Job Morphs to AM

Wondering what your job will really be like in an AM world? See the details in the Indeed.com snippets of actual job openings in Figures 1.4–1.6:

If you're a manufacturing technician in traditional manufacturing, consider the additive roles described in Figure 1.4 or Figure 1.5.

If you're a manufacturing engineer, your AM role might look something like the additive role described in Figure 1.6.

Note that your prior manufacturing experience is expected and valued in the new roles. Also, note that broad AM knowledge, like the content presented in this book, is either expected or preferred.

Additive Manufacturing Design Engineer
Forecast 3D - Carlsbad, CA 92008

Apply Now

- Hands on design support using Solid Works to provide design modifications and suggestions to optimize customer designs for additive technologies.
- Taking potential production programs requirements and being able to transfer the lessons learned from the prototype stage and into production. This includes the development of optimized build for high yield, speed, cost, reduction of hand finishing through the use of DfAM suggestions and process improvement.
- Develop test plans, control plans, DOEs, travelers in support of potential customer production programs.
- Lead and assist with troubleshooting of MJF production optimization to increase efficiency, reduce scrap and rework.
- Ability to evaluate alternative engineering solutions for cost, functionality, strength etc.

Requirements

- Very Strong ability to visualize, sketch and CAD up creative, effective, out of the box design solutions to design challenges
- BA/BS Degree Mechanical Design/Manufacturing/Materials; equivalent experience will be accepted in lieu of a degree
- Master of SolidWorks and other CAD platforms
- Experience with 3D Printers and or within the Additive Manufacturing industry preferred
- Knowledge and experience with Generative design a major plus
- Experience using process improvement tools such as FMEA, PPAP, APQP, Control Plans, IQ/OQ/PQ a plus
- Knowledge of or experience working in ISO9001, AS9100, or ISO13485 environments.
- Six Sigma green belt or black belt a plus

Figure 1.3 Additive Manufacturing Design Engineer Job Description
Source: Excerpt from Indeed.com (accessed 5 November 2020).

Additive Manufacturing Technician

Tesla ★★★⯪☆ 4,519 reviews - Sparks, NV 89434

Apply On Company Site

Tesla is seeking a highly motivated individual interested in acquiring, and expanding, the skill sets necessary to operate SLA/SLS/FDM 3D printing equipment in our rapidly growing Additive Manufacturing operations.

KEY RESPONSIBILITIES

- Perform basic functions, with guidance, required to set up builds on various systems
- Perform all functions related to machine turnaround and preparation for SLS, SLA, and FDM
- Facility clean-up
- Stock parts and materials

REQUIREMENTS

- Passionate about rapid prototyping and 3D Printing
- Must be self-motivated and able to manage multiple parallel-path projects.
- Meticulous attention to detail
- Well organized and strong verbal communication skills
- Flexible, ability to work in a fast pace constantly changing environment with long hours and tight deadlines, including week-ends and off-hours when required
- Independent and demonstrated an ability to deliver successful designs under tight timing constraints
- Basic computer skills including Microsoft Word, Excel, and Outlook

DESIRED SKILLS AND QUALIFICATIONS

- Verifiable experience in a 3D printing work environment
- Customer service / interaction experience
- GD&T and the ability to think in 3 dimensions
- Experience with CATIA, SolidWorks, Pro E, Rhino, or other CAD manipulation software
- 2 year degree or technical certification a plus
- Willingness to work overtime and/or weekends as required, with little to no advance notice, and sometimes for extended periods

Figure 1.4 AM Technician at Tesla

Source: Excerpt from Indeed.com (accessed 5 November 2020).

Additive Manufacturing Technician

Milwaukee Electric Tool Corporation ★★★★☆ 379 reviews -
Brookfield, WI 53005

Apply On Company Site

×

Job Summary:

Duties and Responsibilities

- The primary role of this position is to create 3D Printed components using the latest software and Additive Manufacturing Machinery, in support of New Product Development Teams.
- Work directly from CAD Models to produce accurate 3D Printed components in a variety of polymers and metals.
- Duties to include machinery setup, post processing, maintaining a clean work environment, ordering of raw materials and managing new and completed work requests.
- This position requires a highly motivated, organized individual with good communication skills who will communicate frequently with the New Product Development Teams.
- Maintain schedules in conjunction with safety and quality in a fast-paced environment.

Education and Experience Requirements

- Requires a high school education or equivalent, supplemented with engineering or technical related Additive Manufacturing experience.
- Must have a thorough understanding of 3D printing platforms and materials
- Working knowledge of CAD, and operational knowledge of one or more of the following software systems: Materialise Magics 3D Sprint, 3D Lightyear and Grab Cad.
- A thorough understanding of the following: SLS, SLA FDM, DMLS

Figure 1.5 AM Technician at Milwaukee Electric Tool

Source: Excerpt from Indeed.com (accessed 5 November 2020).

Additive Molding Manufacturing Engineer
Arris Composites - Berkeley, CA

Apply Now

As our Additive Molding Manufacturing Engineer, You Will:

- Create and optimize the process of using composite materials to create end-use parts
- Select the proper methods to achieve the target specifications (e.g. mechanical, thermal, dimensional, surface finish, etc.)
- Determine the processing conditions needed to achieve target specifications (e.g. mechanical, thermal, dimensional, surface finish, etc.)
- Be responsible for hands-on prototyping, fabrication, testing of new molding processes
- Work on simple electromechanical design and assembly
- Manage exquisite record keeping and ability to statistically deduce information from data
- Manufacture workflow and process improvement
- Work collaboratively with technical and non-technical team members to achieve shared goals
- Relay complex engineering decisions to a team of technicians
- Estimate technical risk and time to complete tasks
- Work with a multidisciplinary team in a dynamic work environment

You Will Ideally Bring:

- Minimum BS in mechanical engineering, industrial engineering, operations research, manufacturing engineering, materials science or similar
- 5+ years experience working in manufacturing
- Experience in bonding, welding, melting, and processing of thermoplastic or thermoset composite materials
- Experience in correlating out-of-specification parts to process, material, and/or design issues
- Experience in controlling thermal contraction and expansion of thermoplastic composites during processing
- Experience with impregnating and wetting out fibers with thermoplastic resin
- Knowledge of composite laminate theory and optimal fiber alignments
- Excellent attention to detail
- Good written and verbal communication skills. You can sketch well enough to communicate complex ideas
- Vision and can respectfully disagree when you see things differently than others but you also can recognize when your concept isn't the one that we're moving forward with, and you can get 100 percent behind the new direction
- Great professional references
- Portfolio links welcomed and encouraged

Figure 1.6 Additive Mold Manufacturing Engineer at Arris Composites
Source: Excerpt from Indeed.com (accessed 5 November 2020).

Case Conclusion: Developing Knowledge and Skills in AM

After the team at Great West Manufacturing (GWM) completed their research assignments to investigate additive manufacturing and identify how the traditional manufacturing jobs at GWM might change due to moving into AM, they met to review their key findings and suggestions. The executive summary they prepared contained the following key points:

- All key traditional manufacturing jobs will require employees to acquire new knowledge and skills for GWM to move into AM successfully. Some employees nearing retirement may not be interested in moving into AM, or more recent hires may already have some educational background in AM. Everyone at GWM needs to understand the new manufacturing paradigm and shift away from the traditional manufacturing mindset. Various positions will require more in-depth AM knowledge, but all will need a strong foundation.

- Design engineers, manufacturing technologists, and machine operators will experience the most significant level of change in their job roles. Vendor-specific training for technicians and operators is needed. Design engineers will need to explore new design processes and programs to address the complexity enabled by AM. AM introduces new concepts to design and materials engineers. Part design and build process parameters drive material properties more so than in any other manufacturing process. Understanding the technologies and material selection allows the team to leverage the AM process both as a final product and as a tool in aiding the existing or potential new techniques that are derived.

- Manufacturing engineers and industrial production managers will need to plan carefully to incorporate AM in existing production processes or replace existing methods.

- The manufacturing technician's role requires a deeper understanding of software since the CAD inputs and processing play a crucial role in the system's functioning. In AM, there is variability within the machine and from machine-to-machine.

- Part design and build process parameters drive material properties, a potentially new concept for the design and material engineers. Understanding the technologies and material selection allows the team to leverage the AM process both as a final product or as a tool in aiding the existing or potential new techniques that are derived.

- Testing, compliance, and quality control will be impacted by the need for new processes to evaluate complex parts and gain needed certifications.

- Better communication and collaboration will need to be instilled across the organization to ensure a tight connection between design engineers, manufacturing engineers, and technologists, technicians, and operators.

- At a minimum, the team needs to have a common understanding of additive manufacturing technologies to provide a neutral basis for comparison to their existing processes. Given the compression of the manufacturing and production time frame and the tight connection among job roles and responsibilities, ensuring all job roles are familiar with the complete end-to-end additive manufacturing processes is critical to GWM's success. This comprehensive picture will also help employees ensure that AM makes value-added contributions to the current line of manufactured goods. The goal is to innovate and expand on manufacturing operations at GWM.

HOW TO USE THIS BOOK

This book, written for the student, technician, designer, engineer, manager, and other manufacturing practitioners interested in acquiring and applying additive manufacturing technologies and processes within their organization, aims to provide a broad foundation for ongoing learning. It covers all aspects of AM from product design to production and incorporates additive manufacturing technologies in practical, useful, real-world case studies. The case study further illustrates alternative manufacturing technologies and demonstrates additive manufacturing as the preferred method. Chapter 3 describes the AM design process and inputs to AM. Chapter 4 covers the range of materials used in AM, while Chapter 5 presents the technology and methods appropriate for different materials. Chapters 6 and 7 discuss the preparation for the AM build, the build, and related quality factors. Chapter 8 explores a range of AM post-processing activities, including limited coverage of compliance and testing.

As a manufacturing practitioner navigating this book, you will be part of the journey in analyzing opportunities for the application of AM in your organization. Begin by identifying a problem or opportunity for which you would like to explore an AM solution in Chapter 2. You will work through the major decision steps and processes needed to move an AM initiative forward in subsequent chapters. The case study will explain how another organization has addressed challenges and decisions to move AM forward in their organizations. By the end of the book, you should be prepared to make a case for implementing an initial AM project.

REFERENCES

Dyndrite.com (2020). *HP'S Universal Build Manager Powered by Dyndrite*. www.dyndrite. com/ubm (accessed 9 November 2020).

Grand View Research (2020). 3D printing market worth $35.38 billion by 2027|CAGR: 14.6%. www.grandviewresearch.com/press-release/global-3d-printing-market (accessed 17 May 2020).

Giffi, C., Wellener, P., Dollar, B., et al. (2018). The jobs are here, but where are the people? Deloitte Insights. www2.deloitte.com/us/en/insights/industry/manufacturing/manufacturing-skills-gap-study.html (accessed 17 May 2020).

Sachs, E., Haggerty, J., Cima, M. and Williams, P. (1993). *Three-Dimensional Printing Techniques*. 5,204,055A.

Wellener, P., Dollar, B., Ashton Manolian, H., et al. (2020). The future of work in manufacturing. Deloitte Insights. www2.deloitte.com/us/en/insights/industry/manufacturing/future-of-work-manufacturing-jobs-in-digital-era.html?nc=1 (accessed 17 May 2020).

Chapter 2

Is Additive Manufacturing the Right Solution?

Case Introduction: Selecting an AM Pilot Project

In the Great West Manufacturing (GWM) main conference room, Pete Granger (manufacturing process engineer), Edgar Remmins (manufacturing technician), and Roxanne Jensen (compliance, testing, and quality control engineer) were growing impatient. Bob Nelson (design and materials engineer) was more than 10 minutes late for an important meeting of their group, responsible for developing a proposal to initiate an additive manufacturing pilot project at GWM. At 15 minutes past the meeting time, Bob walked in with a big smile and placed a six-pack of Heineken Lager in the center of the conference table.

BOB: Sorry, I'm a bit late. I wanted to bring some inspiration to the meeting.

PETE: Uh. . .I think we're here to plan the proposal for the AM pilot project. You're not trying to tell us that you can 3D print a beverage, are you?

EDGAR: With the advances they're making in 3D printing, I wouldn't be surprised!

BOB: Very funny! I researched and discovered that Heineken did a one-year pilot project and produced some pretty impressive results. You wouldn't think a brewing company could gain much from AM. Still, their out-of-the-box

Fundamentals of Additive Manfacturing for the Practitioner, First Edition. Sheku Kamara and Kathy S. Faggiani © 2021 John Wiley & Sons, Inc. Published 2021 by John Wiley & Sons, Inc.

thinking on how to apply AM across their entire manufacturing process has resulted in an 80% reduction in time for delivery of required parts to their production line, as well as an 80% cost savings for those parts (Van de Staak, 2019).

ROXIE: I'm all for out-of-the-box thinking, as long as AM can address our needs! Let's learn more about Heineken's approach and explore what other options exist for how we can pilot test AM at GWM! Also, I think we should add someone from business or accounting to the team. We'll need to be able to justify our project financially, and I'm sure they can help.

The team agreed to take an in-depth look at how Heineken applied AM in its plants and explore approaches to piloting AM. They also decided to explore more of AM's advantages and benefits over traditional methods and identify a broad range of AM applications across multiple organizations. They hoped to select and finalize an application for the focus of the pilot project.

INTRODUCTION

AM, initially used for rapid prototyping and limited, customized production, has expanded to become a general-purpose manufacturing technology used for a seemingly limitless range of value-adding applications in the manufacturing process. Current applications cover a wide range. Re-engineering multiple parts into a single component, such as the GE Catalyst engine, which went from 855 parts to 12, resulting in a 5% weight reduction and 20% lower fuel usage, is one example (GE, 2019). Bioprinting human ears based on 3D scans to ensure an identical match for an existing ear is another example (University of Wollongang, 2019). The variability in technologies, materials, and possible applications, while exciting, creates a challenging decision for starting in AM. Teams like GWM's need to make sure their first project doesn't become a paperweight on the CEO's desk but instead demonstrates the value-add of AM in their unique manufacturing environment and produces an immediate economic benefit.

The best place to start in the process of selecting an AM pilot project is with a well-defined business need. By examining the end-to-end manufacturing process, obvious needs may surface. For example, are there significant bottlenecks or time delays in the manufacturing process caused by the inability to quickly replace equipment parts? Could these parts be 3D printed on-site? Can some products be redesigned from using complex assemblies to using fewer parts? Does the company receive requests for product customization that they cannot fulfill cost-effectively? Where is money being wasted in the production process, and is AM a possible solution?

Heineken, a company that operates over 150 breweries worldwide, conducted a one-year test of AM by installing a 3D printer in one of its plants and using a team approach to explore possible applications. The initial business goal was to find a way to source parts for plant equipment to reduce delivery time and part costs. The team quickly defined three initial projects (AM Chronicle, 2019):

1. *Safety latches.* Since worker safety was of crucial importance at the plant, the team started with a simple project to produce plastic safety latches for equipment that would keep machines from being turned on until maintenance completion. Latches 3D printed from a red filament that was highly visible and fit on almost every device in the plant were installed.

2. *Conveyor-belt sensor.* The team then explored the capabilities of 3D printing to redesign and replace broken or damaged metal parts. Engineers were quickly able to create, test, and tweak part designs to obtain the desired properties and functionality. For example, the redesign and 3D printing of a conveyor-belt sensor prevented bottles from jamming up and halting the production line.

3. *Custom maintenance tool.* The third application involved designing and creating a custom tool to allow workers to more quickly and efficiently tighten or loosen the guiding wheels to apply bottle caps. Overall, Heineken reports that by using AM across various applications, they have achieved cost reductions of 70–90% and a decrease in delivery time for the AM applications of 70–90% (AM Chronicle, 2019). As a result of their initial trial period, Heineken disseminated its AM knowledge to all plants and are now exploring how AM can address other needs.

Examining the end-to-end production process can help highlight business needs that indicate possible applications for AM. Having more in-depth knowledge of the benefits of various AM applications achieved in other companies can help evaluate and select from among multiple opportunities. In this chapter, you will learn about the unique characteristics of AM, a wide variety of possible AM applications, the value that AM can add to manufacturing operations, and how to begin to build a business case for incorporating AM in current operations.

ADDITIVE MANUFACTURING APPLICATIONS

Computer-aided design (CAD), introduced in the late 1950s, enabled designers and engineers to use the software in manufacturing processes. The output from 2D CAD files required extensive programming of CNC machines to produce end-use parts from the drawings. Initially, the designs were limited to two-dimensional (2D) drawings until the introduction of Pro/Engineer by PTC in the late 1980s, which supported three-dimensional (3D) model output. The output from the 3D CAD software required minimal programming to produce the final part for machining.

The ease and speed of transition from design to physical part using 3D CAD likely led to the initial emphasis on AM for rapid prototyping. From initial concept to final design approval in the manufacturing process could be significantly shortened. AM provided a means of quickly fabricating parts for design verification and approval, with little attention given to how AM could revolutionize design itself or add value across the product development process.

AM Notable 2.1

AM in Healthcare – 3D Printing Nasopharyngeal Test Swabs

The two companies that manufactured most of the world's nasal swabs for testing during the COVID-19 pandemic were unable to ramp up production rapidly to meet growing demand worldwide. Traditional manufacturing techniques involving injection molding require significant tooling and many additional molds to meet demand. A consortium of additive manufacturers quickly stepped in to help meet this critical need. Origin, a San Francisco-based 3D printing group, quickly designed, sought FDA approval, and conducted clinical trials on a swab approved in record time. Origin had the capacity to 3D print 1500 swabs per hour on each of its 50 machines using programmable photopolymerization (P^3) and medical-grade polymers. It could also easily share the FDA-approved design file with other additive manufacturers, who could quickly ramp up production. In addition to swabs, additive manufacturers could help resolve the personal protective equipment (PPE) shortage by producing glasses, masks, face shields, and ventilator splitters.

Source: Additive Manufacturing Media (2020). How test swabs became 3D printing's production win: The Cool Parts Show, Special Episode. Transcript. www.youtube .com/watch?v=82tLx8AZxSc&utm_source =ActiveCampaign&utm_medium =email&utm_content=Cool+Parts+Bonus+Ep isodes+Available&utm_campaign=Cool+Par ts+Bonus+Episodes. (accessed 09 July 2020).

If viewed as a general-purpose tool instead of a part fabricator, AM technology can fully leverage AM in existing and new processes. As illustrated in the Heineken projects, this thought process allows the designer, manufacturer, student, engineer, and others to develop innovative solutions using AM. In traditional design for manufacturing (DFM), manufacturing professionals focus on a design's ease of manufacturability and cost-effectiveness rather than creating an optimum solution for a given problem. In AM, the reverse happens, with the design focus on developing the optimum solution since there are far fewer manufacturing barriers. Thus, to leverage AM, the optimum solution should be the focus of the design effort, knowing that AM capabilities can address most issues that would arise in manufacturing.

The results from finite element analysis (FEA), a standard process in optimizing engineering design, are more accurately implemented in additive manufacturing than in traditional methods where significant modifications are required to accommodate the manufacturing challenges. For example, in 2018, Bugatti produced a brake caliper, the world's largest functional component made from titanium on a 3D printer.

In designing the brake caliper of a supercar, key considerations included: the high-temperature environment where heat resistance was a priority; high strength to bring the car to a stop from speeds exceeding 200 mph; and light-weighting to improve performance. Traditionally, aluminum was the preferred material due to its ease of machining; however, 80% of the material was wasted in the original design. Titanium, which is optimum for this application considering its strength-to-weight ratio and high melting point, led to a higher cost in traditional machining for the same design due to its hardness. Titanium did not become a viable material for this specific application until the use of AM.

Using AM technology, specifically powder bed fusion in this example, there was no difference in the manufacturing process for aluminum or titanium. The layer-wise buildup of the layers was geometry and material independent. Bugatti utilized an aerospace-grade titanium alloy, which, when tested repeatedly, withstood temperatures exceeding 1857 °F (1014 °C) when nearing the maximum speed of the latest Bugatti Chiron (420 km/h, 261 mph) (Jackson, 2019). The Bugatti brake caliper, printed in titanium, is shown in Figure 2.1.

The Bugatti brake caliper story provides an example of the shift from the traditional "design for manufacturing" thinking to a "design for additive manufacturing" mindset that will help manufacturing professionals fully leverage AM technologies.

A shift in mindset is also needed to move away from the "AM for parts fabrication" perspective to consider a full range of applications that address the entire design and manufacturing process. Table 2.1 summarizes the key AM application areas and describes the advantages AM provides in each of these areas over traditional manufacturing.

Applications of AM may vary widely by industry. The following applications are explored in this section:

- Design verification
- Functional models for validation
- Replacement parts and low-volume manufacturing
- Jigs and fixtures
- Full production and mainstream manufacturing

Specific business needs, AM solutions, and the associated outcomes for the AM applications are provided.

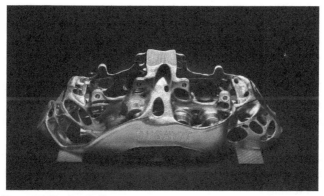

Figure 2.1 Bugatti Brake Caliper
Source: © BUGATTI AUTOMOBILES S.A.S.

Table 2.1 Advantages of AM over Traditional Manufacturing

Industry	Applications	Benefits
Aerospace and Defense	• Functional prototypes • Tooling • Lightweight components	• Low-volume production for complex parts • Material efficiency through reduced waste • Part consolidation through integration of multiple parts
Automotive	• Prototyping • Tooling	• Faster product development • Greater design flexibility • More complex geometries • Customization ease and cost-effectiveness
Industrial Goods	• Production of components • Tooling • Machine maintenance and replacement parts	• Greater design complexity • Shorter lead times • On-demand production reducing costs
Consumer Goods	• Protypes and models for new products • Product testing • Production	• Enhanced product development and testing • Reduced time to market • Mass customization
Medical and Dental	• Prototyping • Patient-specific solutions • Medical devices • Bioprinting	• Enhanced medical device design • Personalized health care • Faster delivery to customer • Reduced production costs

Source: Adapted from Attaran, 2017.

Design Verification Models: Innovative Toy Designs

3D CAD models enabled designers to verify their designs quickly. Prior model creation methods involved machining, wood carving, or other crude methods for producing a physical prototype. These methods were time-consuming and required days or weeks to build a physical prototype. AM changed that by creating physical prototypes in a few hours. Designers were now able to produce multiple variations of their designs within hours to reduce the design time. Parts were accurate enough to fit in an assembly of other parts for further testing.

Figure 2.2 Giochi Preziosi Toys
Source: Photo Credit: Stratasys

In today's increasingly competitive business environment, reduction in time-to-market for products can be critical. Creating a product's physical prototype helps the design and marketing teams finalize designs and move them forward into production. Giochi Preziosi, an Italian toy manufacturer, wanted to evaluate new toy ideas using "true to life" 3D models created for their internal approval process.

Before 3D printers, the models would have been carved or machined in sections, then painted. Traditionally, in the toy industry, photo-realistic prototype models already created by 3D printers would still require hand painting to achieve the required color scheme and detail. The hand painting frequently resulted in inconsistent colors and appearances of toy models. Using AM's material deposition techniques, specifically, the Stratasys® J750™ 3D printer, a full-color, multi-material 3D printer that supports over half a million color combinations, the team at Giochi Preziosi was able to produce photorealistic 3D models with no other painting (Stratasys, 2020a). The AM technology created the models with vivid, accurate, consistent, and repeatable colors, previously unattainable. The toys appear in Figure 2.2.

The ability to accurately produce toy models to support the internal approval process and to easily and quickly modify the designs as desired allowed the Giochi Preziosi team to cut internal approval times by 30%, leading to a reduction in time to market with innovative new toys (Stratasys, 2020a).

Functional Models for Validation

As the portfolio of 3D printed materials and their mechanical properties improved, AM functional models were routinely used to validate designs before moving to production. In cases where the material of choice is paramount, 3D patterns are used for room-temperature vulcanizing (RTV) molds or to produce molds used in injection molding machines to generate the production material of choice for final testing and validation. The Accurate Clear Epoxy Solid (ACES) tool on the stereolithography machine was one of the first processes used to produce engineered plastics using injection molding. In other cases, secondary applications such as plating were used to increase the mechanical properties by multiple factors to validate processes.

Lockheed Martin's Space Systems Company (SSC) division serves commercial, civil, and national security customers with a host of advanced technology systems. In one project, Lockheed needed to validate physical prototypes of satellite fuel tanks. Form and fit were among the first uses of AM design models, which evolved to functional testing as materials improved, and the systems' build volumes increased.

Traditionally, Lockheed Martin's SSC division would have required a lead time of over six months to machine the fuel tanks of 3.8 × 3.8 × 3.8 ft and 6.75 × 3.8 × 3.8 ft at the cost of over $250,000. Due to the large size of the fuel tanks, they could be 3D printed in sections and bonded together using a traditional manufacturing process.

Considering the processing requirements and project deadline, Lockheed partnered with Stratasys Direct Manufacturing, a service bureau. The team utilized material extrusion techniques using a Stratasys Fortus 900mc machine and polycarbonate material to produce 10 sections of the larger tank and 6 smaller tank sections. Each section required around 150 hours to build, but Stratasys Direct Manufacturing's capacity supported multiple Fortus 900mc machine concurrent use (Stratasys, 2020b). After printing, joining tank sections, finishing surfaces, and machining critical areas, the tanks were delivered. The AM approach took a fraction of the time required to produce the tanks with traditional manufacturing.

The tanks then went through several measurements for quality assurance and accuracy and were approved for first-concept assembly. Lockheed Martin's Space Systems Company performed form, fit, and function testing and process development to validate the proposed design changes.

Replacement Parts/Low-Volume Manufacturing: US Navy Compression Pumps

Producing a replacement part to match the original design intent is a challenge that is more pronounced when the original drawings or CAD files are unavailable. With advancements in 3D scanning technologies, it is easy to convert parts to their digital equivalents with intricate detail accurately. The challenge then shifts to producing these components, typically time-sensitive and limited to one or two copies.

The Naval Undersea Warfare Center (NUWC) in Keyport, Washington, needed a replacement vacuum cone casting used on Ohio-class submarines. Using conventional sand casting, the original equipment manufacturer (OEM) quoted $29,562 per vacuum cone, with a 51-week lead time. This approach required the development of a pattern to produce the sand molds for casting the part.

In most sand casting applications, AM significantly reduces lead-time while producing complex castings that are currently a challenge with traditional processes. NUWC was able to reverse engineer the castings to create a 3D CAD file. Using ExOne's AM binder jetting technologies, the team could directly print the sand molds, eliminating the pattern development process. The castings were produced and delivered within eight weeks for only $18,200 per casting (ExOne, 2014). The amount represented a cost savings of 40% per submarine and a 43-week reduction in lead time.

Jigs and Fixtures: TIG Welding Fixtures

Jigs and fixtures are custom-made tools used routinely in manufacturing to guide a cutting tool, hold or support a component, or serve a particular purpose in a manufacturing process. As the complexity of components increases, the design and manufacture of these fixtures is a challenge. Traditionally, fixtures are machined from metals or a combination with other materials, including composites and polymers. AM design freedom enables the manufacture of complicated jigs and fixtures to reduce cycle time and improve ergonomics for the assembler. Increasing the hatch spacing in material extrusion systems produces parts that meet the mechanical property requirement and are lighter, designed ergonomically to reduce fatigue, and increase throughput.

Considering the limitations in traditional manufacturing processes, developing a fixture could require multiple parts in a final assembly. On the other hand, superior designs are possible that are enabled by additive manufacturing. The widespread use of 3D scanning allows designs that conform to the geometry of the actual part. As parts deviate from the original CAD dimensions but are still within tolerance, 3D scanning produces 3D printed files for use as go/no-go gauges.

An aerospace company contracted with the Rapid Application Group, a full-production AM service bureau, to optimize and manufacture welding fixtures for TIG welding. The aerospace company's goal was to reduce the time and cost of fixture production compared to traditional machining.

First, topology optimization software produced complex designs to meet fixture requirements and simultaneously use less material and take less time for fixture production. The optimized designs would have been impossible to create using traditional methods and reduced fixture weight with greater or equal strength. The Rapid Application Group team used AM's powder bed fusion technology, specifically selective laser sintering (SLS) and DuraForm® GF material, to produce the fixtures (3D Systems, 2020). This heat-resistant nylon-based material met the stiffness and temperature requirements to support TIG welding. The finished fixture appears in Figure 2.3.

It took 84% less time to produce the welding fixture, and the cost was reduced by 56%, both of which addressed the aerospace company's business goals.

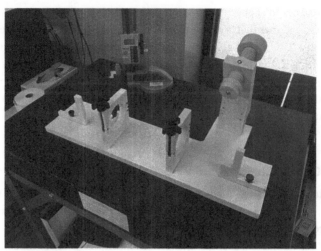

Figure 2.3 TIG Welding Fixture
Source: Courtesy of 3D Systems, Inc.

Full Production and Mainstream Manufacturing: Adidas Midsoles

In instances where the production volumes are less than 5,000 parts, AM has been used to produce parts and avoid expensive tooling costs. Automated finishing systems have improved the surface finish of AM parts, and in some cases, improved their mechanical properties. Plating AM parts with nickel and other metals with increased throughput has enabled batch processing of custom designs.

Moving AM into mainstream manufacturing has been a slow process but has been accomplished by innovative companies with targeted business needs. Adidas partnered with Carbon, Inc. to achieve their objective of creating a one-piece midsole to meet the performance and comfort requirements of serious runners. Also, Adidas wanted to adjust the cushioning properties of the midsole to deliver customized footwear on demand. Traditional manufacturing methods could not produce this type of intricate, high-performance one-piece design that typically required the assembly of multiple components to achieve the desired goal, which is costly and raises quality issues.

Using AM's vat photopolymerization techniques, specifically Carbon's Digital Light Synthesis™ process, enabled Adidas to achieve its business goals and innovate the footwear manufacturing process (Carbon, 2019). Its efforts resulted in the Futurecraft 4D, which uses a midsole similar to that shown in Figure 2.4. AM technology allows Adidas to create a one-piece midsole to meet runners' needs for movement, cushioning, stability, and comfort. The technology also facilitated more than 50 design iterations as Adidas sought to achieve an optimal solution. The number of iterations represents a substantial increase in the number of possible designs considered in the same period using traditional molding.

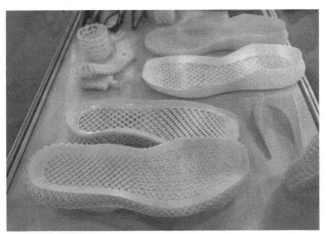

Figure 2.4 Example of 3D Printed Midsole.
Source: asharkyu/Shutterstock

The midsole used a stiff elastomer made from a blend of UV curable resin and polyurethane printed in a lattice structure. The lattice structure, impossible to produce with traditional manufacturing processes, allows for a high-performance midsole with excellent durability and strong aesthetics. The lattice can also be adjusted for different runners to help meet Adidas's ultimate goal of producing bespoke athletic footwear to facilitate mass customization.

Adidas, working in tandem with Carbon, innovated in design, material, and manufacturing processes in a time frame that was not possible in traditional manufacturing. It took 11 months from the first meeting of Adidas and Carbon to Futurecraft 4D's release. The typical time from design to release of a pair of sneakers via average factory production would take 15–18 months (Cheng, 2018). Adidas scaled production quickly, manufacturing the Futurecraft 4D at two of their facilities that were newly equipped with Carbon 3D printers.

Full Production and Mainstream Manufacturing: MEIDAI Eyewear

MEIDAI Technology, a fashion-forward tech startup, initially used traditional injection molding to create noncustomized eyewear. Because of the number of steps required in the manufacturing process and the need to create and maintain between 10,000 and 20,000 different molds, the costs rapidly escalated, quality issues arose, and the production approach became unsustainable. The company also needed an optimal production method to support a new business idea to mass-produce customizable eyewear that customers could design using a Smartphone application.

MEIDAI partnered with Infinite 3D, an AM and 3D printing service bureau, to help meet its business needs to produce customized eyewear. Using MEIDAI's Smartphone application, customers scan their faces and use the software to explore different geometries, designs, and colors. After selecting their customized frames, the customers pay and submit the order. The models are loaded into the software and printed overnight, making them ready for shipment and delivery the next day. Infinite 3D uses HP multijet fusion AM technologies and a thermoplastic material that meets eyewear applications requirements.

AM enabled MEIDAI to implement its business strategy by meeting the need for reduced production cost and supporting product customization via their software without sacrificing quality. Both design and production times were cut for MEIDAI when it moved to AM. Traditional design in the injection molding process typically takes

Table 2.2 AM Applications and Benefits by Industry

Industry	Applications	Benefits
Aerospace and Defense	• Functional prototypes • Tooling • Lightweight components • Jigs and fixtures	• Low-volume production for complex parts • Material efficiency through reduced waste • Part consolidation through integration of multiple parts
Automotive	• Prototyping • Tooling • Limited production • Custom manufacturing • Jigs and fixtures	• Faster product development • Greater design flexibility • More complex geometries • Customization ease and cost-effectiveness • Personalization and unique designs
Industrial Goods	• Production of components • Tooling • Machine maintenance and replacement parts	• Greater design complexity • Shorter lead times • On-demand production reducing costs
Consumer Goods	• Protypes and models for new products • Product testing • Production	• Enhanced product development and testing • Reduced time to market • Mass customization
Medical and Dental	• Prototyping • Patient-specific solutions • Medical devices • Bioprinting	• Enhanced medical device design • Personalized health care • Faster delivery to customer • Reduced production costs

Source: Developed with content from Industrial Applications of 3D Printing – The Ultimate Guide (n.d.).

3–4 days, while the 3D design for AM is possible in 2 days. Infinite 3D can produce 3000 parts in 5 days, whereas traditional injection molding could take 45–60 days to make the same number of pieces (HP Development, 2019). AM services provided by Infinite 3D enabled customer customization of the design and style and a custom fit.

As AM applications have expanded beyond prototyping, its applicability to various industries and the ability to solve design and manufacturing problems has become evident. Time and cost savings from these applications are possible across the entire manufacturing spectrum. Table 2.2 provides a sample of the applications and benefits of AM accrued by industry.

AM Notable 2.2

AM in Fashion – The Future of Jewelry

Two law students made a dramatic career shift when they discovered a business idea made possible by AM. While searching for a duplicate ring they purchased on vacation, Casey and Janine Melvin identified the lack of options for purchasing affordable, customizable jewelry pieces online. Their solution? Their solution was to start TFOJ – The Future of Jewelry. TFOJ is an online jewelry startup using AM to fulfill customer desires.

Artisans have traditionally created custom jewelry at a high cost to consumers. The process involves designing the piece in wax and then using the wax model as a mold for precious molten metals. TFOJ uses software that allows online customers to create their rings based on templates or original works, then transmits the design file to a 3D service bureau that prints the wax model. Online customer design submission saves significant time and cost, allowing the custom piece delivery in six weeks or less. The final cost of the custom jewelry is transparently based on material cost. Prices begin around $65 for a customized signet ring.

Source: The Future of Jewelry (2020). https://thefutureofjewelry.com/. (accessed 09 July 2020).

A BRIEF HISTORY OF AM

Figure 2.5 briefly summarizes key dates in the development of AM, several of which are discussed below.

The initial concept behind AM appeared over 40 years ago when Hideo Kodama, a member of the Nagoya Industrial Research Institute, published information describing the production of a solid printed model. Several years later, Charles Hull patented a method of creating models by curing liquid photopolymer resin using UV lasers, known as Stereolithography (SLA). He subsequently commercialized the first 3D printing system. Around the same time as SLA appeared, Scott Crump developed fused deposition modeling (FDM), which he patented in 1989, and formed Stratasys Inc.

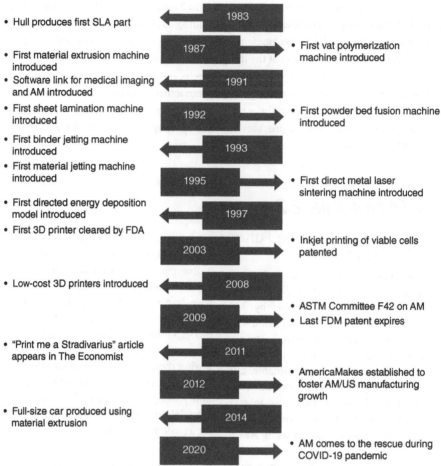

- Hull produces first SLA part — **1983**
- First material extrusion machine introduced — **1987** — First vat polymerization machine introduced
- Software link for medical imaging and AM introduced — **1991**
- First sheet lamination machine introduced — **1992** — First powder bed fusion machine introduced
- First binder jetting machine introduced — **1993**
- First material jetting machine introduced — **1995** — First direct metal laser sintering machine introduced
- First directed energy deposition model introduced — **1997**
- First 3D printer cleared by FDA — **2003** — Inkjet printing of viable cells patented
- Low-cost 3D printers introduced — **2008**
- **2009** — ASTM Committee F42 on AM / Last FDM patent expires
- "Print me a Stradivarius" article appears in The Economist — **2011**
- **2012** — AmericaMakes established to foster AM/US manufacturing growth
- Full-size car produced using material extrusion — **2014**
- **2020** — AM comes to the rescue during COVID-19 pandemic

Figure 2.5 Key Developments in Additive Manufacturing.
Source: Original, Kamara and Faggiani, 2020.

In the late 1990s, the FDA cleared the first 3D printer for medical applications. Scientists 3D printed synthetic scaffolds to attempt to grow a human bladder. In the early 2000s, advancements in 3D printing led to the first commercially viable Selective Laser Sintering (SLS) machine and the founding of Objet. Objet later created a device that could mix multiple materials in the same build. The launch of MakerBot, a low-cost FDM desktop machine, followed and caused massive growth in 3D printing as a hobby (Wohlers and Gornet, 2014).

The publication in *The Economist* of an article titled "Print Me a Stradivarius" propelled 3D printing and AM into the mainstream. It attracted significant capital investment in the technology and applications that continue to promote its growth in manufacturing. New interest in 3D printing accompanied the start of the coronavirus pandemic in 2020. Commercial organizations and hobbyists stepped up to find ways to meet the need for personal protective equipment (PPE) using their printers.

AM growth and development will continue into the foreseeable future as a central element of Industry 4.0. As new materials are invented, printing speeds increase, hardware costs drop, new and innovative applications are discovered, and autonomous production capabilities come online, AM will continue to evolve in the factories of tomorrow.

SELECTING A PILOT AM PROJECT

Manufacturing companies have found many ways to experiment with and incorporate additive manufacturing (AM) into their operations. Selecting and implementing an initial AM pilot project has been the approach used by most companies, regardless of time, budget, organization knowledge, or personnel restraints. Not only does a pilot project allow manufacturing professionals to explore the possible incorporation of AM in their operations, but it provides hands-on training for those involved in the pilot. New knowledge can be quickly acquired and shared across the organization.

Figure 2.6 summarizes the different approaches to piloting AM into the four broad categories. The categories are described briefly in this section. Also, several factors that positively influence the success of an AM pilot project are discussed.

Identify a Pain Point

One way to identify a value-add focus for an AM pilot project is to identify a pain point within the manufacturing process that can be addressed by an AM application. It can be as simple as Heineken creating a plastic part to ensure worker safety during machine maintenance or identifying replacement parts that could easily be 3D printed to shorten lead times and avoid storage costs.

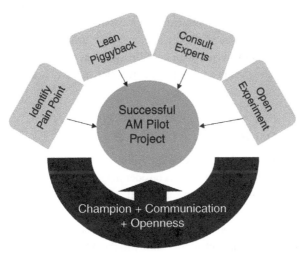

Figure 2.6 A Successful AM Pilot Project
Source: Original, Kamara and Faggiani, 2020.

Questions to consider when identifying a pain point to address with AM:

- Where are we likely to encounter unexpected repairs or unplanned downtime with our equipment?
- What are the high costs of maintenance that plague us? Can these be traced back to specific parts or assemblies?
- Are there safety hazards we have difficulty addressing?
- What are the most common complaints we receive from customers about the performance or maintenance of our products?
- From concept to delivery, where do we spend the most time?
- Where are our highest costs, aside from labor, in our manufacturing process?
- In what areas are we challenged to comply with industry and regulatory standards?
- Where do we experience problems with part or product quality?
- What in our manufacturing process is holding us back from innovating?
- Where are we challenged to manage inventory levels to avoid outage or downtime?

With candidate pain points identified, the next questions are: How can AM help us lessen the pain? And what do we expect to see as the result of applying AM to this pain point?

Involve a broad cross-section of workers across the organization in this exercise to ensure the identification of important pain points occurs, and there is buy-in to the possible AM solution. It is also essential to quantify AM's desired outcome to resolve the issue, such as the expected reduction in lead times, reduction in material waste, reduction in customer complaints, etc.

You will need to ensure that the AM solution is well-bounded and will not necessitate any significant shift in overall manufacturing processes. Since this is the first attempt at exploring AM in the operation, you'll also want to avoid "mission-critical" applications. By selecting a pain point to address with a well-scoped AM pilot project, you can ensure that your efforts will be a value-add for the organization and will not end up as a paper-weight on someone's desk.

Piggyback on a Lean Initiative

Many manufacturing firms have active Lean initiatives in progress at any given point in time. Related to the pain point approach, consider piggybacking on an identified problem or issue for which a Lean team seeks a solution. It may also be possible to bring a pain point to a Lean team and ask them to partner with the AM pilot initiative to define a Lean/AM project jointly. The joint approach ties the AM pilot to direct business goal achievement and the Lean project adds visibility to the AM effort. The AM and Lean partnership is further discussed in Chapter 9 as a method for implementing AM across the organization.

Team Up with Experts

University AM consortiums, AM equipment vendors, and AM service bureaus can provide expertise to assist your organization with identifying and carrying out an AM pilot project. This path is particularly advantageous. It leverages outside expertise to address the manufacturing organization's unique problems and provides an approach to cost-effective manufacture of parts or products during the pilot project.

CEO Jamie Howard of Ultimaker AMERICAS, the manufacturer of AM equipment used by Heineken, suggests a manufacturing facility walk-through. Reviewing the entire production line with additive application engineers, plant managers, plant engineers, and 3D printer operators can identify parts that could benefit from 3D printing. It's not unusual for such teams to find 10–30 applications per plant that could be 3D printed with an accompanying demonstrable business case (Hitch 2019). Service bureaus can provide similar expertise, plus the added benefit of multiple materials and types of AM technologies versus a single vendor. University consortiums, such as the Rapid Prototyping Consortium at Milwaukee School of Engineering, can provide many of the same knowledge and services with the added benefit of custom design expertise and a broad spectrum of experience.

Foster an Experimental Atmosphere

A method of introducing AM within manufacturing firms is making a 3D printer available on the shop floor for experimentation and initial trials by interested parties. When Heineken's packaging manager worked to reduce the costs of producing new parts for plant equipment, he decided to try 3D printing. His cost-reduction quest led to the purchase and installation of Ultimaker equipment and a team's formation to identify applications.

The availability of an AM machine on-site enables manufacturing professionals to gain hands-on knowledge and experience while identifying useful applications. Since AM machine cost has been declining, this has become a feasible approach for small to medium-sized manufacturing organizations. It may also be possible to lease equipment for the pilot project to reduce costs further. By having the AM capabilities readily available, the resources are there when problems arise, and the likelihood that AM will produce a value-added application for the organization increases. It also supports the use of AM capabilities by design engineers for more traditional design and verification activities.

Factors That Influence Success

No matter how an AM pilot project is identified and selected, several factors are essential to its success. These include identifying an AM champion that can advocate for the effort, maintaining transparency and communication to the entire organization about the effort, and ensuring an atmosphere that fosters experimentation and tolerates failure exists.

An AM champion is essential to a successful pilot project. In addition to advocating for resources and assisting in a company-wide communication effort, the champion removes roadblocks that can stall the pilot project and works to build ongoing support for AM within the organization. The ideal champion is passionate about AM and 3D printing, believes the technology can positively impact the organization and its success, and holds a respected position both up and down the organizational hierarchy.

Communicating the purpose, goals, progress, and results of the AM pilot project requires special care and transparency within the organization. AM may represent a significant change for some organizations. Those workers not directly involved may experience anxiety and feelings of job insecurity. A solid communication plan can address worker needs and smooth the transition to AM. Communication can also help generate excitement about AM technology and processes and encourage workers to get involved with the pilot and learn more about AM to prepare for job opportunities that arise with wider-scale implementation.

Since AM will be new to the manufacturing environment during the pilot test, those involved are likely to have limited knowledge. Encourage experiments and expect mistakes. Develop a culture of openness and plan to discuss mistakes made during the pilot process to learn and avoid similar mistakes in future efforts. Errors may arise from choosing an application that is not a good fit for a unique manufacturing environment, poor initial design, incorrect choice of materials, or wrong technology choice. Appropriately scoping the pilot project can minimize the downside of mistakes, and manufacturing professionals need to be encouraged to try again, without penalty. The overall pilot project budget should contain contingencies to address the experimental nature of the AM pilot.

Building a Business Case for AM

A good AM proposal should directly address the anticipated positive impacts of a proposed application by clearly identifying metrics that will demonstrate AM's contribution. The advantages of AM by application appear in Table 2.1, and the AM benefits by industry appear in Table 2.2. These advantages and benefits provide the initial factors to help formulate a business case for a proposal. Almost all the projected positive impacts are quantifiable as cost savings, cost avoidance, reduced material waste, reduced downtime, reduced inventory and shipping costs, and similar metrics. A strong partnership between manufacturing and business professionals can help identify and quantify a pilot project's AM impacts. Evidence-based AM success will justify extending AM implementation within the company.

AM Notable 2.3

AM in Food – "Real food, freshly printed"

AM and 3D printing are finding their way into kitchens and generating new culinary creations and stimulating creativity, such as the 3D printed guacamole lizard appetizers. Natural Machines and their Foodini 3D printer originated, of course, from a business need. When a vegan bakery wanted to expand beyond country borders, the owner discovered that manufacturing and distribution accounted for 80% of production costs. In comparison, raw ingredients and labor accounted for only 20% of costs.

The solution? A 3D food printer that functions as a mini-manufacturing kitchen appliance and enables the chef to be a creator, maker, and manufacturer all in one. The Foodini can print multiple foods by pushing them down a capsule and through a nozzle and accurately layering them into stacked-layer, three-dimensional food designs. The machine comes loaded with preset shapes and creations, or the cook can easily create original food works.

3D printed food restaurants are rapidly appearing, such as Food Ink in London, where all food, utensils, and furniture are 3D printed to provide an immersive, futuristic experience for diners. Make your reservations at FoodInk.io!

Source: Retrieved from www.naturalmachines .com/how-it-works (accessed 09 July 2020.

Case Conclusion: Identifying AM Applications for a Pilot Project

The first thing the Great West Manufacturing (GWM) team did was add a new member to provide expertise on the business side of manufacturing operations. It was easier than they thought it would be to find an enthusiastic member to join the effort. Frank Burns, finance and accounting manager, volunteered to join the team when approached for suggestions about who might be interested in his department. His daughter Emma, a sophomore in an engineering program at a local university, had purchased a personal 3D printer for her dorm room last summer. Frank and Emma spent time together working on several exciting projects before the start of the Fall term. He was intrigued with additive manufacturing and believed it could significantly impact GWM's bottom line.

Adding Frank's new perspective, the team designed and implemented a mixed approach to piloting the technology to help involve as many GWM employees in the effort as they could. They hoped to build interest and energy in the organization around AM. Their process included the following steps:

- Acquire an inexpensive AM machine, primarily for use by design and manufacturing engineers. With an AM machine available, engineers can gain familiarity with the AM design process from concept to prototype and start moving their mindset away from traditional manufacturing design to design for AM.

- With the GWM CEO's approval, the team held a contest to identify AM applications for the pilot project. Winners of the competition would receive an additional day off with pay, in addition to their standard benefits package. A web-based, online training seminar on AM basics was available to all employees to educate them on how AM works and its applications. After viewing the session, each department discussed challenges or opportunities in their respective areas to identify one AM application that could make a difference within their department or across departments.

- Suggestions from the departments had to meet several criteria. In short, any suggested AM application had to:
  - Be doable within a three- to six-month time frame.
  - Have a high likelihood of positively impacting at least one of the following, as measured by a metric selected by the department and approved by the GWM AM Team:
    - Reducing product or operating costs
    - Reducing waste or material costs
    - Improving customer satisfaction and the associated likelihood of repeat sales
    - Reducing downtime for manufacturing equipment/processes
    - Simplifying highly complex product or parts
    - Increasing the safety of employees
    - Other outcomes as approved by the GWM AM team

    Not wanting to hamper creativity and innovation by the department employees, this was the only criteria provided. Within the given requirements, employees were able to identify the need they wished to fill and formulate an AM application to address it.
- A review panel was formed to evaluate the AM application suggestions. The AM team decided to augment their AM knowledge by bringing in some experts to help assess the applications suggested by GWM departments. The evaluation team consisted of:
  - AM team members
  - Two application engineers, one specializing in metals and the other in polymers – both of whom previously worked for AM equipment vendors, from a 3D printing service bureau located in a neighboring city
  - The director of the AM Institute at a state university
  - GWM's CEO, Sherman Potter

  The review panel evaluated the suggestions and selected the top three AM applications for inclusion in the pilot project. As review panel members, the service bureau engineers helped garner favorable prices for any contract 3D printing work needed as part of the pilot project. The AM Institute was willing to provide additional training and design support for GWM during the pilot project, as required.
- The review panel received a total of 11 AM application suggestions from the departments at GWM. After careful evaluation and consideration, the group chose three applications to move forward into the GWM AM Pilot Project. The projects selected were:

- *Customer Service, in conjunction with Design Engineering.* Redesign of a multipart component that generates a significant number of customer complaints due to the difficulty in assembly; the project focus is a new design with fewer parts that's easier and quicker to assemble, as enabled by AM. Metrics: Reduction in customer support requests or complaints related to the component.

- *Manufacturing Press Operators.* Develop on-demand printing of several critical components of manufacturing equipment – unexpected breakdowns cause unplanned downtime with a negative impact on productivity because replacement parts are difficult to source. Metrics: Reduction in downtime; reduction in repair cost

- *Cost Accounting and Purchasing Department.* Replacement of several metal parts with a redesigned polymer part printed in-house to reduce overall product cost and improve the needed part's supply chain availability. Metrics: Reduction in product cost; reduction in lead time on part orders

Many outstanding AM applications were suggested but not included in the pilot project. CEO Potter made clear the priority given to the applications not included in the pilot and set up a tentative schedule to address them in the coming AM expansion.

REFERENCES

3D Systems. (2020). Production time, cost for TIG welding fixtures reduced with SLS 3D printing. www.3dsystems.com/customer-stories/production-time-cost-tig-welding-fixtures-reduced-sls-3d-printing (accessed 05 July 2020).

AM Chronicle. (2019). Heineken utilizes Ultimaker S5 3D printer to develop custom tool and functional machine parts. www.amchronicle.com/news/heineken-utilizes-ultimaker-s5-3d-printer-to-develop-custom-tool-and-functional-machine-parts/ (accessed 06 June 2020).

Attaran, M. (2017). The rise of 3-D printing: The advantages of additive manufacturing over traditional manufacturing. www.researchgate.net/publication/313904803_The_rise_of_3-D_printingThe_advantages_of_additive_manufacturing_over_traditional_manufacturing/link/5955368a0f7e9b591cd7391e/download (accessed 09 July 2020).

Carbon, Inc. (2019). Carbon lattice innovation – the Adidas story. www.carbon3d.com/white-papers/carbon-lattice-innovation-the-adidas-story/ (accessed 05 July 2020).

Cheng, A. (2018). How Adidas plans to bring 3D printing to the masses. *Forbes* (May 22). www.forbes.com/sites/andriacheng/2018/05/22/with-adidas-3d-printing-may-finally-see-its-mass-retail-potential/#568df18d4a60. (accessed 5 July 2020).

ExOne (2014). US Navy Compression Pump. www.exone.com/en-US/case-studies/
US-Navy-Compresson-Pumps (accessed 05 July 23020).

GE (2019). Disrupt. . .or be disrupted: Why 3D printing is changing the world of
industrial design and manufacturing. www.pressreleasefinder.com/prdocs/2019/
GE_Additive_Overview_May2019.pdf (accessed 06 June 2020).

Hitch, J. (2019). Additive success brewing on factory floor. *Industry Week* (May 24). www
.industryweek.com/technology-and-iiot/article/22027651/additive-success-brewing-
on-factory-floor (accessed 15 June 2020).

HP Development. (2019). Customized eyewear made possible with MEIDAI and HP
multi jet fusion technology. Case Study. https://enable.hp.com/us-en-3dprint-meidai
(accessed 05 July 2020.)

'Industrial Applications of 3D Printing: The Ultimate Guide' (no date) AMFG. Available
at: https://amfg.ai/industrial-applications-of-3d-printing-the-ultimate-guide/
(accessed: 14 December 2020).

Jackson, B. (2019). Watch: Bugatti prepares 3D printed brake caliper for series production. https://3dprintingindustry.com/news/watch-bugatti-prepares-3d-printed-brake-
caliper-for-series-production-146163/ (accessed: 05 July 2020).

Stratasys. (2020a). Giochi Preziosi uses Stratasy J750 to 3D print innovative toy designs.
Case Study. file:///C:/Users/kfagg/Downloads/CS_PJ_CN_GiochiPreziosi_EMEA_
A4_EN_0120_Web.pdf (accessed 05 July 2020).

Stratasys. (2020b). Lockheed Martin 3D prints fuel tank simulation with help
from Stratasys Direct Manufacturing. Case Study. www.stratasysdirect.com/
resources/case-studies/3d-printed-fuel-tank-lockheed-martin-space-systems-
company?resources=d6b6b549-daef-40a9-bd8d-e19d5cb1d80e (accessed 09 July 2020).

University of Wollongang (2019). '3D Alek' multi-material bioprinter makes human
ears. Medical Products Outsourcing. www.mpo-mag.com/contents/view_breaking-
news/2019-02-28/3d-alek-multi-material-bioprinter-makes-human-ears/ (accessed
06 June 2020).

Van de Staak (2019). Heineken: Ensuring production continuity with 3D printing. https://
ultimaker.com/learn/heineken-ensuring-production-continuity-with-3d-printing
(accessed 06 June 2020).

Wohlers, T., and Gornet, T. (2014). History of additive manufacturing. Wohlers Report.
Wohlers Associates, Inc., pp. 1–3. www.wohlersassociates.com/history2014.pdf
(accessed 9 November 2020).

Chapter 3

What Design and Inputs Does Additive Manufacturing Require?

Case Introduction: Sourcing Inputs

After reviewing the three recommended AM pilot applications, CEO Potter and the AM Pilot Project team decided to address the part redesign suggested by the Customer Service and Design Engineering groups. Their original submission involved a complex multipart component that customers were required to assemble. Consumers were dissatisfied with the complicated assembly process, leading to numerous calls to the customer support group.

The component selected for the pilot project is a quartz burner system used in a small-profile, high-end gas cookstove favored by campers. The quartz burner system

Fundamentals of Additive Manfacturing for the Practitioner, First Edition. Sheku Kamara and Kathy S. Faggiani © 2021 John Wiley & Sons, Inc. Published 2021 by John Wiley & Sons, Inc.

currently consists of 16 different parts preassembled into five sub-components that the consumer assembles into the final cookstove after purchase.

The AM Pilot Project team added representatives from the customer service and design engineering groups. The next steps involved redesigning the original multipart assembly with a design for additive manufacturing (DFAM) mindset. The team knew they needed to produce a digital file of the new design as an input to the AM process. The new digital design file posed a challenge since there was no digital representation of the existing production parts. The design engineers also felt it would be valuable and help them gain knowledge of AM capabilities if they first recreated the existing parts using AM. Potter and the team agreed, so they set out to complete the following next steps:

1. Explore their options and select a method for creating digital files of the existing parts.

2. Using the newly created digital files, generate the file format needed to perform the 3D printing process.

3. Have an AM service bureau 3D print the parts.

4. Test and evaluate the 3D printed parts.

5. Using their new AM knowledge, redesign the multipart component using DFAM to take advantage of AM capabilities and address the desired metrics for reducing component complexity and customer dissatisfaction.

6. Generate the file format needed for AM machines.

The team agreed that it needed to perform these tasks in a time-efficient matter while also allowing for new learning about the AM process and sharing the information with the rest of the organization along the way.

INTRODUCTION

After selecting an application for AM, design takes center stage. The application may involve creating a brand-new part or product, modifying an existing part to improve it, or recreating an existing part with AM. Whatever the application, the goal in this step of the process is to produce a 3D digital representation of the part that can be formatted and fed into an AM machine.

This chapter discusses the differences between traditional manufacturing design and design for additive manufacturing (DFAM), describes AM design processes, and presents the hallmarks of a good AM design with a design evaluation worksheet suitable for novice to intermediate AM designers. The next step in AM design is creating a file in a format readable by AM machines. Making the most popular file formats for AM machines using various CAD software packages or other technologies is then described. The chapter concludes with the latest update from GWM's AM Pilot team.

DESIGN FOR AM

Traditional manufacturing design focuses on design for manufacturability, design for assembly, or combining the two. Design for manufacturability keeps costs as low as possible while considering the optimization of parts manufacturing in early design. For example, designing dies for plastic injection molding requires taking care to design draft angles that allow for part removal from the mold or allowances needed to adjust for shrinkage. Design for assembly also seeks to keep costs down but pushes ease of assembly into early design considerations. For example, if two parts need joining, fasteners are a standard solution; however, fasteners make up an average of 5% of materials cost and up to 70% of labor costs. AM's new manufacturing methods and technologies allow designers to focus on optimizing the part design itself, with sophisticated software that helps address the manufacturing and assembly issues to help reduce costs (Keane, 2016).

Moving from Traditional Design to Design for AM

Moving from a traditional manufacturing design mindset to DFAM requires discarding the old design rules. In AM, there are far fewer constraints than conventional techniques, allowing the designer to focus more fully on optimizing the part to meet requirements. While the old design approaches may work and effectively integrate with AM design processes, it's important to sit back, get creative, and ask: What can 3D printing do that wasn't possible before? Using AM advantages, how can I add value to the part or product?

Despite the lack of manufacturing constraints in AM, other important considerations exist. Different materials, methods, and machines can produce dramatically different results considering requirements and cost models that may need careful evaluation during the design process.

Validation and finalization of a design would be impossible without this information. The following design discussion describes design tasks common to all AM materials and techniques. Chapter 4 discusses available AM materials and their properties, and Chapter 5 presents methods and technologies plus potential contributions of different materials and processes to part requirements.

DESIGN FOR DFAM

From a design engineer's perspective, AM design knowledge consists of three areas: component design, part design, and process design (Wiberg, Persson, and Olvander, 2019). Each of these categories consists of several steps, as illustrated in Table 3.1. Tasks related to component and part design are emphasized in the following discussion to describe key practical design considerations for those new to AM.

Table 3.1 DFAM Overview

Component Design	Select a Component	Identify the right product or component for AM:
		• Types of components suitable for AM
		• Characteristics of components suitable for AM
		• Type of value AM adds to components
		• Business case with combined cost model
	Define the Design Problem	Identify boundaries between components and their interfaces:
		• Integrate functional design to reduce parts and add value
		• Focus on functionality while taking full advantage of AM capabilities
	Define Requirements and Constraints	Identify requirements and constraints for the component to be manufactured:
		• Analyze the design problem using traditional engineering techniques
		• Requires evaluation of AM materials, processes, and cost models
Part Design	Create an Initial Design	Determine what the part should look like:
		• Optimize structural components using topology optimization (TO) formulations/commercial software tools that address AM constraints
		• Apply AM advantages to standard features in nonstructural cases
	Interpret the Initial Design	Interpret and adapt initial design for AM, as needed:
		• Apply design rules specific to AM method, material, machine type, and settings
		• Consider AM restrictions and features impossible with other manufacturing methods
	Verify the Design	Analyze using CAE and verify structural, thermal, and other properties:
		• Design complexity and anisotropic material in AM are unique
		• AM part surface texture depend on geometry and build direction, making ability to simulate properties difficult

Table 3.1 *(Continued)*

Process Design	**Create and Evaluate Support Structures**	**Analyze geometry and build direction to identify support needs:** • Determine if support structures needed • Optimize shape of part and placement of support structures • Optimize build direction to minimize need for support structures • Change part design to be self-supporting
	Prepare for AM	**Set up the machine prior to manufacturing:** • Establish machine settings appropriate to part, method/material, and software used • Settings may include energy-related, scan-related, powder-related, and temperature-related depending on part characteristics
	Validate Build and Cost Time	**Prepare a cost model for the part or component:** • Determine build time as basis for cost model, including material, # of components in a build, etc. • Optimize cost based on design and production technique
	Simulate the AM Process	**Simulate the AM process to observe productivity, surface quality, part dimensions and other qualities:** • Use commercial simulation software to determine part quality, such as: Netfabb, Siemens NX, Materialise Magics, Simplify 3D, 3D Systems, 3Dsim, or Simufact Additive

Source: Content adapted from Wiberg, Persson, and Olvander, 2014.

COMPONENT DESIGN

Component design, also referred to as system design, takes a broader view of a given portion of a product and attempts to answer the following questions:

1. Is AM an appropriate manufacturing method for this component?

2. What are the part boundaries and connections within the component selected?

3. What requirements must the part meet (tensile strength, load, thermal properties, etc.)?

A two-step process can determine if AM is an appropriate method for producing a component. First, sort candidate components into one of the following groups (Klahn, Leutenecker, and Meboldt, 2014):

A. Using AM is likely to be beneficial.

B. Using AM could be beneficial, but further evaluation is necessary.

C. Using AM is likely to produce no benefits.

D. Using AM is not appropriate.

Components in groups C and D are unlikely to benefit from AM. For those components in groups A and B, the next step is to consider the following benefits or value-adding possibilities of AM, which include:

1. Functional integration of design and improvements in functionality results in a reduction in the number of parts required to achieve component requirements and improve customer experience.

2. Individualization or mass customization potential for the component results from economies of smaller batch size production.

3. Component weight and material usage are reduced due to AM materials or manufacturing process.

4. Greater design efficiency – AM enables more complexity without generating additional cost.

The decision about whether AM is suitable is facilitated by first identifying the likelihood that benefits will accrue and by examining the benefits likely to be achieved.

When defining the design problem, the focus should be on identifying the part boundaries and interfaces to integrate the design to reduce the number of parts and associated assembly costs. The edges will likely change within the component, while interfaces to other system components are maintained. A design method for reducing the number of parts, which integrates some traditional engineering design practices with DFAM, is illustrated in Figure 3.1.

DFAM methods are still evolving, and more comprehensive approaches are under investigation. Several AM design researchers suggest developing automated design systems that provide a framework in which AM design guidelines are systematically applied, especially by AM novices. One such framework involves an additive manufacturing database of rules integrated with axiomatic theory and inventive problem-solving to generate new and innovative solutions for design problems (Renjith, Park, and Kremer, 2019).

The third set of component design activities focuses on identifying the requirements and manufacturing constraints for AM components. Many techniques for analyzing design problems and specifying requirements in traditional manufacturing processes can be applied to AM, though manufacturing rules for AM are dependent on the AM method and possibly by the specific AM machine used. For example, the maximum build size depends on each AM machine. AM materials also play a significant role, as most are anisotropic; thus, mechanical properties and surface textures depend on build orientation and AM method.

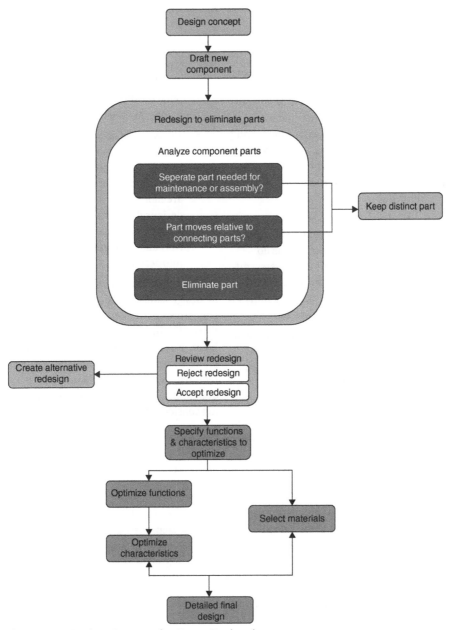

Figure 3.1 Design Process for Parts Reduction
Source: Adapted from Rodrigue, H., & Rivette, M. 2010.

Although AM constraints are few compared to most traditional approaches, they can be challenging to specify unless AM materials, processes, and technology are selected simultaneously. Equipment vendors who retain their proprietary information may not make this information readily available. Those new to AM may find it valuable to work with an AM service bureau with additive application specialists familiar with multiple materials and machines or a university center with comparable knowledge.

PART DESIGN

Given part requirements and manufacturing constraints, the parts design tasks involve creating, interpreting, and verifying the design. These steps are essential to AM design before conversion to a digital form that can be read by the 3D printing process.

Many different design approaches and techniques are available. Three areas, highly compatible with AM, are presented here:

1. AM design guidelines
2. Topology optimization
3. Generative design

AM design guidelines derived from equipment vendors, experience, and researchers' findings have become a part of standard practice. Topology optimization and generative design techniques optimize AM designs within a set of boundary requirements using algorithms and artificial intelligence. Generative design methods are made possible by AM. Topology optimization focuses on determining the least material usage in a given set of requirements. These processes produce designs manufacturable using AM with minimum or no additional modifications. The next sections discuss the three design approaches.

AM Design Guidelines

Experience across various AM applications, materials, and processes has resulted in general guidelines applicable in DFAM. These guidelines are summarized as AM design principles and appear in Table 3.2.

Material, process, and sometimes individual machines strongly influence DFAM and associated constraints. The AM design guidelines presented in Table 3.2 have been generalized and apply to most AM technologies.

Table 3.2 AM Design Principles

Topic	Principle Description
Part Orientation	Print orientation can impact surface finish, strength, need for support structures, and cost in terms of number of layers required. Print orientation should be optimized to part requirements, material, and AM process.
Support Removal	Need for support structures should be minimized to reduce costs and save time, as use of supports will impact surface finish and increase need for post-processing. Optimal part orientation can be used to eliminate or reduce support requirements, as can allowing sharp inner edges.
Part Hollowing	Thick walls and hollow interiors can reduce print time and costs.
Manufacturing Constraints	AM can be used to generate prototypes for future manufacturing with traditional processes.
Part Interfaces	AM finite build space may require large parts to be manufactured in smaller sections and later joined together, or multiple parts of a component may be required to meet a specific need; Joined or interlocking gap dimensions should be minimized to enable ease of support removal and ensure smallest possible deviations in dimensions.
Part Count Reduction	AM capabilities enable reduction in number of parts for assemblies or end-use products.
Identification Marks	Use of identification marks on all parts desirable to keep track of parts and models.
Sharp Edges	Sharp edges should be avoided to the extent possible, which can result in better accuracies as rounding radii correlate with outer radii of simple-curved elements.

(Continued)

Table 3.2 *(Continued)*

Topic	Principle Description
Inner Edges	Rounded inner edges simplify removal of support structures and excess materials.
Points	While horizontal, points should be blunted orthogonal to the build plane.
Overhangs	Overhangs should be short to ensure manufacturability and prevent layers from falling off.
Island Position	Low island positions can have a significant impact on build times.

Source: Content adapted from Booth et al., 2016.

Small to medium manufacturing firms may not have invested in topology optimization or generative design software or may not have experienced designers or engineers to apply them in the design process. A more straightforward approach to DFAM, particularly useful for novice to intermediate designers and engineers, was developed at Purdue University (Booth et al., 2016). The method, best used in the early conceptual design stage, involves evaluating a part or component design using eight categories defined on a worksheet. The worksheet uses a scoring scheme to determine the likelihood of the design's AM build success. Figure 3.2. displays the worksheet.

To use the worksheet, the designer or engineer marks the scale for each of the categories (Complexity, Functionality, Material Removal, Unsupported Features, Thin Features, Stress Concentration, Tolerances, and Geometric Exactness). Each mark receives 1 point, points total across each row, and the row multiplier is applied to compute a row total, which are added to produce a total score interpreted as one of four outcomes: needs redesign, consider a redesign, moderate likelihood of success, and higher likelihood of success. The results support a better evaluation of designs as well as a higher probability of 3D printing success. The Design for Additive Manufacturing worksheet, used in a high-volume 3D printing lab, resulted in a 42% decrease in print failure and a 39% decrease in the reprint rate.

Topology Optimization

Topology optimization (TO), a method frequently used for structural components, has evolved to address various AM manufacturing constraints. While TO may be performed manually by experienced design engineers, TO approaches increasingly rely on different optimization formulas and mathematics using sophisticated software to produce the optimized design given manufacturing constraints. The use of software enables less-experienced design engineers to achieve better results.

The designer or engineer first defines the design space in terms of area, volume, loads, and constraints. TO software supports selecting optimization factors or objectives that may include minimizing mass, maximizing stiffness, surface smoothness, and many

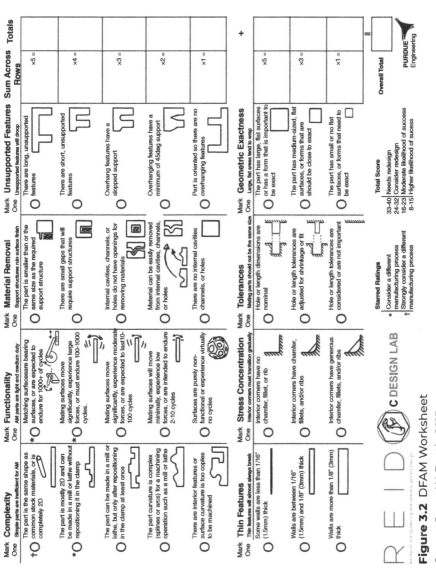

Figure 3.2 DFAM Worksheet

Source: Joran W. Booth, 2015.

others. TO provides an optimized shape based on parameters during the early conceptual design for further refinement. TO may also be used to refine structural components and systems in other phases of the design process.

The design of a roof bracket challenged BMW. The roof bracket needed to raise and lower, plus fold and unfold the soft top on the 2018 BMW i8 Roadster. Casting a device to lift, push, and pull the weight of the roof was impossible. Using TO software, BMW designers input the roof's weight that needed to move and the space available, then generated and refined a design that distributed the load and minimized the amount of material required for the part (Putre, 2018).

The BMW roof bracket designed for the i8 Roadster was 44% lighter than the traditionally manufactured bracket for the previous roadster model. It became the first 3D printed metal component put into full production. It was possible to print more than 600 brackets in a single batch (Putre, 2018).

TO can produce optimized shapes with the following advantages:

- Weight reduction
- Minimized material requirements
- Selection of cost-effective materials
- Analyzing composite materials to deliver optimal mechanical performance
- Reduction in design time
- Higher part or product quality with a lower overall development cost

Software packages that support topology optimization can be tailored to specific applications and processes and provide similar capabilities and functionality. Unfortunately, commercially available software doesn't offer adequate support for AM manufacturing constraints, which necessitates manual conversion of AM results described in the Topology Optimization section. Table 3.3. summarizes a list of the most popular commercially available optimization software tools and their features.

Table 3.3 Topology Optimization Software Tools

Software Package	Overview
Altair OptiStruct	Supports minimum member size, maximum thickness, draw direction, patterning, extrusion constraints, and grouping
Altair Inspire	Offers optimization objectives, stress and displacement constraints, acceleration, gravity, and temperature loading conditions
Vanderplaats Genesis	Allows radial manufacturing constraints to fill material in radial elements as well as symmetry and member thickness constraints
Simulia Tosca	Supports structural and flow optimization based on industry standard finite element analysis packages (AMSYS, MSC, Nastran)
Abaqus ATOM/TOSCA	Capable of fluid flow and heat transfer optimization
MSC Nastran	Finds optimal distribution of material; can generate a conceptual design proposal to be used as initial design for sizing and shape optimization

Table 3.3 *(Continued)*

Software Package	Overview
SolidThinking Inspire	Capable of handling member thickness only; does not support size and shape optimization
Within Enhance	Part of Autodesk; has integrated finite element analysis and can produce organic shapes through optimization
PERMAS-TOPO	Supports conceptual development through specification of design space with variable and fixed pars, boundary conditions, loads, target definition with remaining volume, and additional optimization constraints
FEMTools Optimization	Solves general and structural optimization problems and creates new designs with a layout optimized for load
Optishape-TS	Supports optimization given volume, mass, compliance, displacement, Eigenvalues
Kinematics/Nervous System	Provides 4D printing capability to create complex, articulated, foldable forms; can turn a three-dimensional shape into a flexible structure; used for jewelry and fashion design

Source: Adapted and extended from content in Reddy et al., 2016.

AM Notable 3.2

Designing for Plastics

When creating plastic objects, some design guidelines include:

- Avoid overly thick walls, since they use more material than necessary and may cause parts to warp or deform. If thick walls are needed, sparse internal wall structures can save material and minimize warpage risk. AM process is a key determinant of wall thickness limits for plastics.

- With some processes, parts built layer-by-layer are anisotropic, meaning they have weak points caused by print orientation. Thin external elements are prone to breakage, so part features parallel to the build platform at the bottom of the 3D printer should be avoided. These can be dealt with by either redesign or a change in part orientation on the build platform.

- Some printing processes for plastics produce very high dimensional accuracy and repeatability, such as fused deposition modeling (FDM), which has a standard accuracy of 0.15% with many plastic materials. For plastic parts that interlock or join, or other applications requiring high dimensional accuracy, thermosets on vat photopolymerization systems have higher accuracy and smoother surface finish.

Hopefully, it is becoming evident that design decisions must closely relate to AM material and process. Seeking assistance from material and process experts can help ensure the success of initial experiments with AM.

Generative Design

Generative design is an alternative design approach that supports the exploration of design variants beyond what is possible in traditional design. It uses advanced algorithms to work within stated parameters and goals to generate thousands of design variants. At each iterative design step, structural testing occurs, with new learning incorporated into the design. The process mimics nature's evolutionary forces and takes advantage of technological advances in artificial intelligence and infinite computing. The generative process typically results in designs that would not have been created using traditional processes and have unique, organic shapes to meet specific needs. The lack of manufacturing constraints with AM makes it a good fit for generative design processes. Figure 3.3 illustrates the four main steps of generative design (McKnight, 2017).

General Motors (GM) primarily uses 3D printing for proof of concept. However, they explored generative design and 3D printing with the redesign of a seat bracket. After inputting information about the design space, goals, and constraints, they used generative design software to produce over 150 design options for the seat bracket.

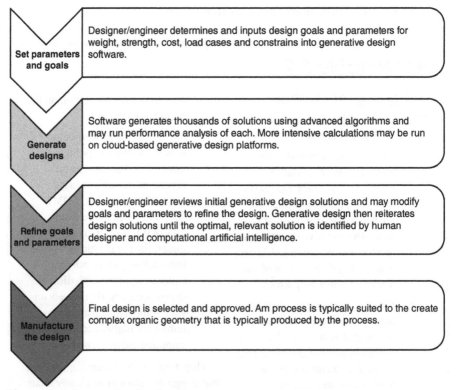

Figure 3.3 Generative Design Process
Source: Adapted from Matthew McKnight 2017.

As a result of applying the generative design process, GM was able to redesign a seat bracket that was 20% stronger and 40% lighter than the bracket currently in use. The approach also enabled part consolidation from eight different components to one 3D printed part (AMFG, 2018).

The possible advantages of using generative design include:

- Quickly identify designs that balance performance and cost.
- Present alternative methods of production.
- Facilitate part consolidation.

Software tools for generative design allow designers and engineers to input various design parameters, including material, size, weight, strength, manufacturing process, and cost constraints. The software uses artificial intelligence algorithms to produce many design alternatives that the designer may not consider. Generative design software presents multiple ways to solve a problem and allows the designer to choose the solution that will work best based on the intense analysis done via the computer.

PROCESS DESIGN

The next set of tasks in the DFAM workstream focuses on the preparation of the design for manufacturing. Broadly defined, these tasks include:

- Creating and evaluating support structures, as needed
- Preparing for the AM build
- Validating build and cost time
- Simulating the AM process

Not all designs or AM manufacturing technologies require support structures. The initial task is to determine whether a support structure(s) is necessary. Since support structures add time and cost to the manufacturing process, the following steps can reduce support needs for a part or component:

1. Optimize the shape and placement of the support structure to avoid the need for supports.

2. Optimize the build direction or orientation to avoid the need for supports.

3. As a last resort, consider redesigning the part to make it self-supporting if the time and cost of support creation exceed desired parameters.

Some topology optimization and generative design software packages provide automated methods to minimize or eliminate the need for supports (Das et al., 2015). Some of the engineering design software that incorporates AM manufacturing constraints also simulate the AM process to provide preliminary results and performance data, such as ANSYS (www.ansys.com/products/structures/additive-manufacturing).

Simulation tools can help conduct design verification, set up build orientation, simulate the 3D printing process, and give a greater understanding of material performance.

After completing the part design steps, including design verification, evaluation of support structure requirements, validation of build time and costs, and AM process simulation, and preparation of the source file, input to the AM process is the next step.

AM Notable 3.3

Designing for Metals

Creating parts with metals involves a different set of design considerations. A few critical metal design guidelines follow:

- Dimensional accuracy can be challenging in indirect metal part production since parts require post-printing sintering to achieve mechanical properties. Sintering produces part shrinkage that is difficult to predict or control. Post-processing to bring a part within tolerances is needed if the printing process alone does not achieve the desired tolerances.

- The software manages the energy density for direct metal part production since thicker areas may increase stresses,

causing part deformation and unstable build processes. Minimum wall thickness is energy-source and material dependent but can range from 0.2 mm to 2 mm.

- During the cooldown of melted powder layers in some AM processes, thermally induced stresses can lead to build failure or part deformation. Designs should incorporate rounded edges and avoid sharp edges and should steer clear of large material accumulations.

In all cases of design for metal objects, designers should be aware of standard tolerances and material properties and how the materials interact with AM processes.

SOURCES OF INPUT

The only input required for producing an additive manufactured part is the digital representation of an object in a 3D file. This requirement provides an avenue for information ranging from reverse engineering to AI-generated 3D models. Almost all engineering design occurs using a CAD system. AM came to reality with the availability of mainstream CAD systems in the late 1980s. A CAD 3D model and AM can produce and assemble complex geometry that required multiple components in traditional manufacturing.

Since AM doesn't require reprogramming a design file, just conversion to another format, design errors will automatically transfer to the physical prototype. Considering computer hardware challenges in the late 1980s and translating a part or component CAD design to a 3D printer, Chuck Hall and 3D Systems developed the STereoLithography (STL) file format. The STL format is a list of triangular surfaces that describe a computer-generated solid model; hence, it is the most straightforward format for creating "slices" of an object for AM processing.

Simulation software produces optimized structures and designs with minimal or no modifications for validation or manufacturing. Human thinking, limited by the proposed manufacturing process or material, may obscure the design intent and optimized design. These limitations affect design thinking, and often designers settle for the design that is suitable for manufacturing. Several sources for the input of 3D files and tools to produce designs for your problem are included in subsequent sections.

AMF AND STL FILE FORMATS

The STL file is the most widely used format in additive manufacturing, notwithstanding its limitations. STL is the *Standard Tessellation Language* or *Standard Triangle Language*. It comes from a list of the triangular (tessellated) surfaces that describe a computer-generated solid model. The term used for breaking the geometry of a surface into a series of small triangles, or other polygons, is *tessellation* (Library of Congress Digital Preservation, 2019).

The solid model is the most accurate representation of the design. The polygon and slice models approximate the solid model; hence, the polygon model's accuracy depends on the parameters used for file conversion.

It is critical to verify the quality of the STL file, as it affects the final part's accuracy. Changing the chord height and accuracy parameters in most CAD systems can minimize the difference between the solid model and the STL files. A larger chord height results in a less accurate model. Figure 3.4 shows a file with standard parameters during translation. Greater chord height depicts lower resolution, fewer triangles, and a less precise file.

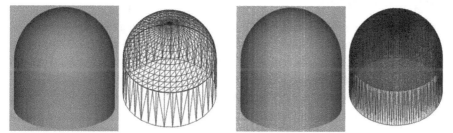

Figure 3.4 STL Format Model with Different Parameters
Source: Original, Kamara, 2020

The file with custom parameters and reduced chord height results in higher resolution and a more accurate file.

Other file formats, such as the AMF (Additive Manufacturing Format) and 3MF (3D Manufacturing Format), address these limitations (All3DP, 2020). The AMF is an ASTM and ISO standard, though it lacks adoption within the industry. 3MF developed through a consortium effort led by Microsoft and major OEMs in the AM industry to speed up adoption (3MF Consortium, 2020).

Both AMF and 3MF are XML-based formats with native support for geometry, scale, color, materials, lattices, duplicates, and orientation.

CAD Systems and Software

CAD software is the primary source for creating AM files for printing. Most designs derive from the start as individual components or assembly for a proposed product. With the wide use of STL files, most CAD software easily translates a design file from its native format to STL. Based on the intended purpose, there are a variety of options that focus on specific industries and applications.

3D Scanning

Producing 3D digital copies of existing parts became more cost-effective with the availability of low-cost 3D scanning systems and software. Most 3D scanning systems with custom software can translate 3D digital models directly to the STL format. Depending on the scanned data, additional software platforms can further manipulate digital information. Mainstream CAD software inputs scanned data to re-create parametric models that can be improved upon while maintaining the original design intent. As AM parts and assembly designs are more organic, using traditional measurement techniques such as calipers and coordinate measuring machines (CMMs) to validate designs are problematic. Contact and non-contact scanning techniques can validate AM components.

Medical Models

With the introduction of the Materialise Mimics medical modeling software in the early 1990s, image data from CT scanners is translated into STL format to produce anatomical models. The segmentation of data allowed for both soft and hard tissue models. The use of these models ranges from surgical preplanning through the development of 3D models that represent unique patient anatomy to custom instrumentation and implants to meet patient needs for surgery.

Other Sources

The simplicity of the STL file format allows for its creation from other sources.

Freeform Surfacing Packages

Software packages such as Geomagic Freeform provide a touch-based haptic interface that allows designers to convert their handcrafting into organic, digital designs. They can add aesthetic and functional details to their plans, handle digital clay, and produce complex voxel-based organic shapes that would not be enabled by traditional CAD systems.

Mathematical Packages

Considering the simplicity of the STL file format, one can output designs from mathematical packages and formulas, such as MATLAB. STL files can also be loaded into MATLAB and modified, then rewritten to STL format.

Photographs

A variety of web-based programs, such as Smoothie-3D, enable users to upload a 2D photograph and convert it to a 3D model for printing. Other tools use multiple digital images to reconstruct a 3D model utilizing 3D scanning software. The software exports the 3D model to STL and other formats that can be 3D printed.

Standard Inputs by AM Technology

Although the STL is the most widely used format in AM machines, some AM technologies may require other file formats to produce the desired output results or models. Models are described by these categories:

- *Monochrome.* A single part or assembly consisting of one material and one color
- *Composite.* A single part or assembly consisting of multiple materials and one color
- *Colored.* A single part or assembly consisting of multiple materials or binders and multiple colors
- *Hybrid.* A single part or assembly produced using a combination of AM and a traditional manufacturing process

Different file formats describe 3D parts or assemblies used in AM processes, listed here:

- *3MF – 3D Manufacturing File.* An XML-based open source file format standard developed by Microsoft and the 3MF Consortium.
- *AMF – Additive Manufacturing File.* An XML-based open source file format developed by ASTM to address shortcomings with the STL file format.
- *OBJ – Object abbreviated.* An ASCII or binary open source file format that stores color information; OBJ is the preferred format for 3D printing in color.
- *G-CODE – Geometry coding language.* Originally developed for programming computer numerical control (CNC) machine instructions, now used to move parts within a 3D printer.

- *M-STL – Multiple STL files.* Used to reference individual part files of an assembly or component.

- *PLY – Polygon File Format.* Also referred to as Standard Triangle Format; stores data from 3D scanners.

- *STL – STereoLithography File.* Also referred to as a Standard Tessellation file; STL is an ASCII or binary format file that is the most widely used file for describing surface geometry of a 3D object.

- *VRML – Virtual Reality Modeling Language.* Represents 3D interactive vector graphics initially designed for the Web but used for 3D printing. WRL file extension used.

- *X3D – Third-generation VRML file; ISO ratified.* An XML-based file format that is an advanced version of the .WRL file format and is an ISO/IEC standard.

Case Conclusion: Design and Inputs to AM

CEO Potter made clear at the outset of the pilot project that GWM would not make a significant investment in AM technologies and processes until the pilot project proved the worth of AM. A clear vision of the future of AM at GWM was required to drive investment in AM. The acquisition would only occur with a tight alignment of AM with business goals and expected returns specified. The AM Pilot Project team determined that the best move forward would be to work with a local university to provide expertise and training to move ahead with the pilot project. The team reached out and negotiated a reasonable cost, acceptable to GWM leadership, for assistance redesigning the selected component and printing the initial prototype. The university AM staff could use their 3D scanner equipment to scan each of the existing quartz burner system's 20 parts and print them using polymer modeling material and binder jetting as an initial proof of concept. The parts were delivered to the AM Pilot project team for evaluation and assembly into the 3D prototype.

The team noted the following findings from the quartz burner prototype process:

- The 3D scanner used by the university easily imported to GWM's current CAD system. The 3D scanner is easy to use for generating a digital model of parts for which no recent CAD files exist.

- AM 3D printing was able to produce the parts for the quartz burner system within acceptable tolerances. However, university staff noted they needed to make some adjustments for several of the part interfaces to ensure a tight fit.

- The GWM CAD system can output an STL file, and since STL is the most frequently used AM format and is compatible with the university AM equipment, GWM will use STL file formats for input into the AM process.

- When the team attempted to do the same assembly required by customers, the reasons for the support phone calls became apparent. It was not as simple a task as they thought to assemble the subcomponents into the burner component.

Working in conjunction with the university AM design engineers, GWM's engineers followed the redesign thought process for part reduction. The resulting design yielded some surprising results. The designers were able to produce a single monolithic part, reducing the number of pieces from 20 to 1 and eliminating customer assembly. The designers were also able to estimate that the redesign would result in a 50% reduction in material volume.

The team couldn't wait to print a working proof of concept prototype, but before they could take the next steps in the pilot project, they needed to make some decisions about AM materials, processes, and technology. The materials would need to have the thermal properties appropriate for the cookstove application and meet safety requirements. The AM technology and process selected would also need to produce a burner system that met all the performance criteria that made this product a customer favorite.

REFERENCES

3MF Consortium (2020). The Linux Foundation projects. 3MF Consortium Membership. https://3mf.io/membership/ (accessed 08 August 2020).

All3DP (2020). 2020 most common 3D printer file formats. https://all3dp.com/1/3d-printer-file-format/ (accessed 08 August 2020).

AMFG (2018). Generative design and 3D printing: The manufacturing of tomorrow. https://amfg.ai/2018/10/25/generative-design-3d-printing-the-manufacturing-of-tomorrow/ (accessed 08 August 2020).

Das, P., Chandran, R., Samant, R. and Anand, S. (2015). Optimum part build orientation in additive manufacturing for minimizing part errors and support structures. *Procedia Manufacturing, Vol 1*: 343–354, doi.org/10.1016/j.promfg.2015.09.041 (accessed 08 August 2020 at http://www.sciencedirect.com/science/article/pii/S2351978915010410).

Keane, P. (2016). What is Design for Additive Manufacturing? *Engineers Rule. Technology for Design and Engineering*. www.engineersrule.com/design-additive-manufacturing/. (accessed 15 July 2020).

Klahn, C., Leutenecker, B. and Meboldt, M., 2014. Design strategies for thr process of additive manufacturing. In: Procedia CIRP. [online] Elsevier, pp.230-235. www.sciencedirect.com/science/article/pii/S2212827115008938 (accessed 9 November 2020).

Library of Congress Digital Preservation (2019). Sustainability of Digital Formats: Planning for Library of Congress Collections. STL (Stereolithography) File Format Family. https://www.loc.gov/preservation/digital/formats/fdd/fdd000504.shtml (accessed 08 August 2020).

McKnight, M. (2017). Generative design: What it is? How is it being used? Why it's a game changer! In *The International Conference on Design and Technology*, KEG, pp. 176–181. DOI 10.18502/keg.v2i2.612.

Putre, J. L. (2018). With a small but mighty bracket, BMW raises the roof on 3-D printing. *Industry Week* (August 7). www.industryweek.com/technology-and-iiot/article/22026127/with-a-small-but-mighty-bracket-bmw-raises-the-roof-on-3d-printing (accessed 08 August 2020).

Renjith, S. C., Park, K., and Okudan Kremer, G.E. (2020). A design framework for additive manufacturing: Integration of additive manufacturing capabilities in the early design process. *Int. J. Precis. Eng. Manuf.* 21, 329–345. doi.org/10.1007/s12541-019-00253-3.

Rodrigue, H., and Rivette, M. (2010). *An Assembly-Level Design for Additive Manufacturing Methodology*. Proceedings of IDMME – Virtual Concept, Bordeaux, France, October 20–22, 2010.

Wiberg, A., Persson, J., and Olvander, J. (2019). Design for additive manufacturing – a review of available design methods and software. *Rapid Prototyping Journal* 25 (6): 1080–1094. doi.org/10.1108/RPJ-10-2018-0262.

Chapter 4

What Materials Does Additive Manufacturing Use?

Case Introduction: Selecting a Material for the AM Application

The GWM AM Pilot Project team completed the part profile and list of requirements that a material, process, and technology must meet to fulfill the new design intentions. The summary of information relevant to material selection is:

- The camp stove is a high performance, ultralight product capable of a high heat output of 12,000 BTU.

- The camp stove used BF1010 butane fuel canisters, and the material selected must be approved for gas conduit applications.

- The material must maintain compliance with the following standard: ANSI Z83.11-2006/CSA 1.8-2006(R2011) Gas Food Service Equipment standards.

Fundamentals of Additive Manfacturing for the Practitioner, First Edition. Sheku Kamara and Kathy S. Faggiani © 2021 John Wiley & Sons, Inc. Published 2021 by John Wiley & Sons, Inc.

- The new material must add no weight overall to the camp stove.
- The new material must be certifiable as a gas conduit.

In the meeting to review the details and plan the next steps, the following conversation occurred:

BOB: I'm starting to understand just how interconnected the AM process and technology are with the material selection. We've done things the traditional way at GWM for so long by designing for our manufacturing process that this is a significant shift! It was great to be free from that limitation, but now we need to deal with practical concerns.

PETE: I hear you on that, Bob. It was exciting to watch the design process unfold with a focus on the optimal solution. Now we just need to focus on the options for the right material. I'm confident there are enough process and technology capabilities out there that at least one will work for us – plus, we can explore all three simultaneously if we bring in some outside AM expertise.

ROXIE: I hope you're right, Pete. From what I understand, the part's functional properties and selected material will vary based on AM process and technology used, and even on build orientation. I've got to make sure what we choose will allow us to meet the ANSI standard target, as well as health and safety requirements. Maybe we can even eliminate the need to place the California Proposition 65 warning on the camp stove. I think sales and marketing would love that!

ED: I was surprised that build orientation could make a difference in material properties, but I guess that makes sense given the layer-by-layer construction process. I'm hoping whatever material we use will be easy to store and work with and safe for me to be around!

FRANK: Good point, Ed! And I'd like to make sure we're not adding a sizable chunk to the cost of our camp stove! (groans heard from the rest of the team). Aw c'mon, you know we've got to stay focused on our business goals!

At the end of the meeting, the team had the initial prototype produced by their University partner, the list of functional and performance requirements, and the standards and certifications the part and end-product needed to meet. They were ready to approach several AM service bureaus for advice and learn more about their application materials.

INTRODUCTION TO AM MATERIALS

From avocados to titanium alloys, thousands of materials are possible in AM and 3D printing, with new and novel materials in development and testing. Polymers are the most common 3D printing material given the broad base of polymer 3D printers installed and used. Metal materials are experiencing high growth driven by the automotive, aerospace, and defense sectors. Metals produce complex, lower cost, high-quality parts that meet stringent performance and safety requirements. Composites, which combine two or more different materials to enhance the resulting material properties, are also experiencing increased demand for use in AM. Elastomeric materials that are soft, flexible, tough, and strong are experiencing growth in the medical and industrial spaces. Although high material costs are a challenge hampering AM for full production, growth in AM materials will likely triple over the next five years (AMFG, 2020).

After discussing material selection and the role of material, technology, and AM design process, the next section discusses materials. Topics include compatible technologies and processes, common applications, advantages, and disadvantages of materials.

SELECTING AM MATERIALS

AM materials are available in one of the following forms: powders, filaments (pellets), liquids, or sheets. In AM applications, melted materials form the layers of an object being 3D printed, though pressurized sheet materials also create objects. For liquids or resins, the solidified materials form the desired item. Each material used requires different 3D printing parameters during the build process with an AM technology, and parts printed with other materials and technologies will have varying properties. In AM, the material change that occurs because of the technology and layering process will determine solid-state changes, final mechanical properties, and design capabilities. AM materials and processes are summarized in Figure 4.1.

Designers should factor in the variability of part properties that results from a material selected, AM technology employed, and the process used after formulating the initial optimal design. The application, performance, aesthetics, and geometry of the object must be considered in the material selection and AM process decisions (Stratasys Direct, 2020).

Application

Material certifications for flame retardance, skin contact, biocompatibility, sterilization, heat smoke toxicity, or other requirements could be required. Both material and AM processes may impact these certifications. Design specifications should identify certifications and other legal and safety requirements to define appropriate materials and AM processes.

Performance

Although function and performance are dependent primarily on design, AM material and process heavily influence the object's final mechanical properties. Structural load, environmental conditions, and other mechanical requirements support the design and material selection decision. Density (D), tensile strength (TS), flexural strength (FS),

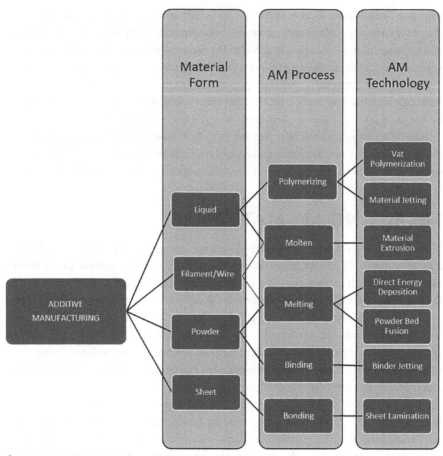

Figure 4.1 AM Materials and Processes
Source: Adapted from Lee, J.-Y., An, J., & Chua, C. K., 2017.

elongation at break (EB), hardness (H), tensile tear resistance (TT), and heat deflection temperature (HDT) are the properties that are the primary focus.

Aesthetics

Surface finish, opacity, and other aesthetic qualities of the object may impact material and AM process selection. Some materials and processes can produce rougher or smoother surface finishes, and post-processing to provide a desired finish or aesthetic may be necessary. The ability to achieve some surfaces may also strongly depend on the AM process employed. Although aesthetics may not play a role in all applications, some materials and techniques can prevent the need for post-processing and should be considered in the selection process to help control time and costs.

Geometry

As stated, different AM materials and processes produce different results. Geometric properties, such as desired wall thicknesses, dimensional tolerances, and minimum feature executions, should factor into AM material and process selection. Design adjustments to address part geometry may be feasible to accommodate an optimal material and process selection. Critical factors for consideration in most materials selection decisions include standard accuracy (SA), layer thickness (LT), minimum wall thickness (MWT), and whether the material can accommodate interlocking or enclosed parts (IE).

Materials, AM technologies, and processes are tightly linked with design in their abilities to produce parts and products that meet application and performance requirements, aesthetic needs, and address unique geometries. Descriptions of materials categories, industry applications, material properties, and AM technologies that process the materials are in the following sections. Detailed explanations of the AM technologies and processes and their impact on materials follow in Chapter 5.

POLYMERS

Polymers, both synthetic and derived from renewable materials, are low-cost, easy to manufacture, versatile, and water-resistant, making them popular for AM across a wide range of products and industry sectors. AM polymers fit into two categories: thermoplastics and thermosets.

Thermoplastics

Thermoplastics are polymers that can be shaped and molded at elevated temperatures but solidify when cooled. Amorphous and crystalline are two categories of thermoplastics. Amorphous thermoplastics have melt characteristics that make them well-suited to material extrusion processes. Amorphous thermoplastics include most clear plastics, such as ABS, polycarbonate, acrylic, and PETG. Semicrystalline thermoplastics perform better for the melting and fusion that occurs in powder bed fusion processes. This material is used in most general plastic parts and includes polyethylenes, polypropylene, nylon, and fluoropolymers.

Several popular amorphous thermoplastics are summarized below (Goldschmidt, 2020):

ABS (Acrylonitrile Butadiene Styrene), available in both filament and powder form, is a widely used 3D printing polymer that is less expensive than other materials. With a melting temperature around 200 °C, ABS requires a heated build plate. It contains a base of elastomers that makes it more flexible and shock-resistant. It also has a longer lifespan compared to polyamide/nylon. ABS appears in automotive bodyworks, mobile phone cases, appliances, and many other applications.

PLA (polylactic acid), one of the most widely used filaments, is a biodegradable polymer formulated from renewal resources such as corn starch and is one of the easiest materials to print. PLA, which does not require a heated print plate or print platform, is used in open-air printing environments. A high cooling and solidification speed, and a slight shrinkage tendency after cooling, make PLA a more difficult material to manipulate. It can print with sharper corners compared to ABS and is available in different colors.

ASA (acrylonitrile styrene acrylate) is like ABS but more challenging to use due to styrene emissions during the printing process. ASA has more excellent resistance to UV rays.

PET (polyethylene terephthalate) is the material most used in disposable plastic bottles. It is the ideal filament for any application requiring contact with food. PET is also more rigid with good chemical resistance. PET's key advantages are that it releases no odors when printing, and it is 100% recyclable.

PC (polycarbonate) is a widely used industrial thermoplastic valued for high strength and transparency in AM. It has a lower density than glass, capitalized on the design of optical parts, protective screens, and similar applications. PC reacts to humidity and must be stored in an airtight container to prevent it from impacting performance.

High-performance polymers (PEEK, PEKK, ULTEM) are advanced polymer filaments with high mechanical and thermal resistance that perform like metals. These are very strong and much lighter than some metals, making them highly suitable for the aerospace, automotive, and medical sectors. High-performance polymers are available in filament and powder form and are compatible with some fused deposition modeling (FDM) machines as well as selective laser sintering (SLS) processes.

Compared to semicrystalline thermoplastics, amorphous thermoplastics show better impact resistance and dimensional stability. However, amorphous thermoplastics are also prone to cracking and have low fatigue resistance. Semicrystalline thermoplastics demonstrate better electrical properties, a lower friction coefficient, and better chemical resistance; however, they are also difficult to bond. They have a sharp melting point and average impact resistance.

Thermosets

Thermosets are thermosetting polymers that behave much differently from thermoplastics. Thermosets do not melt when heated or reform when cooled. Instead, these polymers undergo irreversible chemical bonding during a curing process, which typically involves exposure to light. Such polymers are called photopolymers. AM photopolymers consist of monomers or oligomers and a photoinitiator. Photopolymers may contain additives such as dyes, antioxidants, toughening agents, or others that modify the material's behavior and properties. The AM vat polymerization process cures liquid photopolymer resins under ultraviolet light. Stereolithography (SLA), digital light processing (DLP), material jetting (PolyJet), and continuous liquid interface production (CLIP) are AM processes that also use thermoset materials. These processes also use acrylics and acrylic

Table 4.1 Selected Polymer Material Properties

Material	Application	Performance	Aesthetics	Geometry
ABS	End-use items, jigs or fixtures, concept modeling, prototyping. Biocompatibility (ABSi, ABS-M30).	**D:** 1.05 g/cm³@25°C **TS:** 22 MPa **FS:** 41 MPa **HDT:** 90°C @0.45 MPa 76°C@1.81 MPa **EB:** 6%	• Opaque, multiple colors • Rough finish with fine finishes possible • Can be smoothed, painted, or coated	**SA:** +0.15% **LT:** 0.25 mm **MWT:** 1 mm **IE:** Yes
PC	Engineering environments or applications requiring high flexural and tensile strengths.	**D:** 1.2 g/cm³ **TS:** 68 MPa **FS:** 104 MPa **HDT:** 138°C @0.45 MPa 127°C@1.81 MPa **EB:** 4.8%	• Rough finish with fine finishes possible • Can be smoothed, painted, or coated	**SA:** +0.15% **LT:** 0.18–0.25 mm **MWT:** 1 mm **IE:** No
Polyamide (Nylon)	High mechanical and thermal resistance, chemical resistance, long-term stability, biocompatibility, no support structures needed. Also inexpensive, easily dyes and colored.	**D:** 0.95 g/cm³ **TS:** 48 MPa **FS:** 41 MPa **HDT:** 86°C @1.82 MPa **EB:** 20%	• Grainy surface with fine finishes possible • Can be sandblasted, colored/impregnated, painted, covered, or coated	**SA:** +0.3% **LT:** 0.12 mm **MWT:** 1 mm **IE:** Yes
NeXT Resin	Use when high accuracy and high feature detail needed for tough, complex parts, good moisture and thermal properties. Used for functional prototypes, duct work, connectors, automotive housings and dashboards.	**D:** 1.17 g/cm³ **TS:** 42.2 MPa **FS:** 41 MPa **HDT:** 56°C @0.46 MPa 50°C@1.81 MPa **EB:** 9%	• Building layers visible on surface but high gloss to course finishes possible • Can be sandblasted, painted, varnished, covered, or coated	**SA:** +0.2% **LT:** 0.1 mm **MWT:** 1–3 mm **IE:** No
Xtreme Resin	Good all-around properties ideal for tough enclosures, replacing CNC machined parts, finished models.	**D:** 1.18–1.3 g/cm³ **TS:** 38–44 MPa **FS:** na **HDT:** 62°C @0.45 MPa 54°C@1.82 MPa **EB:** 14–22%	• Building layers visible on surface but high gloss to course finishes possible • Can be sandblasted, painted, varnished, covered, or coated	**SA:** +0.2% **LT:** 0.1 mm **MWT:** 1–3 mm **IE:** No

Source: Derived from Materialise (2020) and Senvol (2019).

hybrids. Photopolymers used in AM must have low-volume shrinkage to avoid distortion of the solid object.

Standard resins used by AM to create parts can produce a high level of detail and a smooth finish, though the range of available colors is limited. It has low shrinkage, high chemical resistance, and is rigid and delicate. Many resin properties are similar to that of ABS. While the surface finish will be good, the mechanical properties will be average. More advanced resins used in dentistry are also biocompatible. Resins formulated for flexibility and nondeformation are ideal for making jewelry. Liquid photopolymers have expanded through research to meet the demand for high-temperature resistance, greater impact, and high elongation properties.

A summary of common polymer materials and their application, performance/ mechanical properties, aesthetics, and geometry characteristics appears in Table 4.1.

Metals

A wide variety of metal materials are available for AM, typically as powders and wire feedstock. The most used powdered metals include aluminum, copper, cobalt chrome, nickel alloys, stainless steels, and titanium alloys. Wire feedstock used includes steel and stainless steel alloys, titanium, tungsten, molybdenum, and aluminum. The variety of metals available is increasing as AM applications take advantage of their availability. Tensile strength (TS), hardness (H), and elongation are factors that go into metal selection decisions. These characteristics are among those summarized for selected metals in Table 4.2.

The process of atomization produces metal powders. In atomization, gas is injected into a molten metal stream before it leaves a nozzle, creating many droplets that freeze before they contact a container surface. Particle size and spherical geometry are the typical specifications for metal powders fed into AM metal processes.

AM metal processes rely predominately on a laser-melting approach to produce layers 10–50 μm (Metal AM, 2020). As previously noted, the material, process, and machine used will impact mechanical properties, and the initial design may be subject to modification to address these potential AM variations. Parts produced by AM, regardless of the metal used, require post-processing to meet specifications or improve surface quality. The estimate of overall manufacturing cost should factor in post-processing costs.

Several categories of AM metals and their applications appear below:

Tool steels are carbon and alloy steels that, as the name suggests, make good tools. These steels have excellent hardness, are abrasion-resistant, are deformation-resistant, and hold their cutting edges at high temperatures.

Stainless steels contain a higher chromium content than carbon steels and resist corrosion, rust, or stain with water. These steels appear in engineering applications in aerospace, oil and gas, food processing, and medical applications.

Titanium and titanium alloys are lightweight and corrosion-resistant and have excellent strength and toughness, making them ideal for aerospace and motor racing

Table 4.2 Selected Metal Material Properties

Material	Application	Performance	Aesthetics	Geometry
Titanium	Popular alloy with low specific weight, corrosive-resistant alloy used in aeronautics; also used for functional proto-types, solid parts, and medical devices due to biocompatibility.	**D:** >4.36 g/cm³ **TS:** >900 MPa **FS:** na **HDT:** na **EB:** >10% **H:** >310 HV	• Rough surface but can be finished for smoothness.	**SA:** na **LT:** 0.03–0.6 mm **MWT:** 1 mm **IE:** No
Stainless steel	Low-carbon alloy with excellent strength and corrosion resistance, good for food-safe applications, parts, production tools, medical instruments, and wearables.	**D:** >7.91 g/cm³ **TS:** >510 MPa **FS:** na **HDT:** na **EB:** >45% **H:** >170 HV	• Rough surface but can be finished for smoothness.	**SA:** na **LT:** 0.03–0.1 mm **MWT:** 1 mm **IE:** No
Aluminum	Low weight, good strength, thermal properties, and flexible post-processing options make alloy suitable for housings, engine parts, production tools and molds in aerospace, automotive, and automation.	**D:** >2.59 g/cm³ **TS:** >250 MPa **FS:** na **HDT:** na **EB:** >1.0% **H:** >80 HV	• Rough surface but can be finished for smoothness.	**SA:** na **LT:** 0.03–0.1 mm **MWT:** 1 mm **IE:** No
Inconel	Nickel-chromium-based alloy suitable for extreme pressure and temperature environments, used in the energy industry for turbines and engine parts or for low-temperature applications such as cryogenics.	**D:** >8.07 g/cm³ **TS:** >940 MPa **FS:** na **HDT:** na **EB:** >8.0% **H:** >300 HV	• Rough surface but can be finished for smoothness.	**SA:** na **LT:** 0.03–0.1 mm **MWT:** 1 mm **IE:** No
Precious metals (gold, silver, platinum – data provided for 14k gold))	Ideal for low-volume, high-end applications to support high customization and design freedom. Jewelry, watchmaking, dental, and electronic applications are most common.	**D:** >12.8 g/cm³ **TS:** 467 MPa **FS:** na **HDT:** na **EB:** 38% **H:** 135 HV	• Rough surface but can be finished for smoothness.	**SA:** na **LT:** na **MWT:** na **IE:** na

Source: Content derived from Senvol (2019) and SpecialChem (2020).

applications. Biocompatible forms of titanium are ideal for biomedical implants or dental applications.

Aluminum alloys have good thermal properties, and their lightweight makes them suitable for thin-walled parts such as heat exchangers, automotive, and aviation parts.

Nickel-based alloys and superalloys combine high tensile, strength, fatigue-corrosion, and thermal fatigue resistance. One such alloy, Inconel, performs very well in extreme conditions at both high and low temperatures, making it ideal for aerospace, chemical processing, and power industry applications.

Cobalt chromium alloys with good biocompatibility, high strength, and outstanding corrosion resistance make good artificial knee and hip joints, engine components, wind turbines, and jewelry.

AM Notable 4.1 - 3D Printing a Ride to Mars

Relativity Space, a SpaceX competitor, recognized that rocket production hasn't changed in 60 years. While a small percentage of rocket parts are now 3D printed, its goal is to 3D print 95% of all rocket parts using super-sized 3D printers. The move to AM with adaptable, scalable, and autonomous robots enabled the company to reduce part count from 100,000+ parts to < 1000 parts, reduce build time from 24 to 2 months, simplify its supply chain, and create a software-defined production environment vs. a traditional operation with high physical complexity.

The materials used? Relativity Space created several proprietary alloys for its Terran 1 rocket and printed at its Stargate factory utilizing the world's largest 3D printers (Relativity Space, 2020). The materials passed all testing for high strength and achieved the physical properties needed for mission-critical structural requirements. Relativity's in-house materials lab quickly iterated new alloy development and designed alloys specifically for its Stargate 3D printers.

Check out an interview with Tim Ellis, CEO at: https://www.youtube.com/watch?v=zeQTrWU1RlU.

Precious Metals

Gold, silver, and platinum, and other precious metals can be 3D printed. These metals' filaments are processed in powder form and generally used in the jewelry sector, though they may appear in electronic components. These metals have high electrical conductance and are heatproof but are expensive to use and difficult to work with because of their high reflectivity and thermal conductivity. Since high temperatures are required, specialized 3D printers are necessary.

Table 4.2 provides a summary of key properties for select metal materials. Additional information about metal AM materials is available in material data sheets available on material and technology vendor websites. In addition to many of the standard materials referenced here, most companies have developed their material offerings that vary widely by technology. Most vendors are willing to offer material selection advice when provided with a detailed description of the application, required performance, aesthetics, and desired geometry of the part or product.

Composites

Material science is a critical factor in the ongoing evolution of additive manufacturing. The growing field of composite materials is at the forefront of development, with the ability to 3D print composite materials viewed as the path to industrial end-use production (Sher, 2017). The drive to manufacture substantial and geometrically complex parts for lightweight aircraft and automobiles prompts AM technology advancement and new composite materials development. Composite materials have also been growing in importance due to the demand for lightweight materials with high strength for specific applications. Polymer composites and metal composites are discussed in the next sections. Table 4.3 provides a summary of key properties for select composite materials.

Polymer Composites

A polymer composite is a polymer material reinforced with fillers from another material, usually fibers of natural or synthetic materials. Hybrid fiber-reinforced composites are polymers combined with two or more fillers (Rajak et al., 2019). Natural fibers used in AM include hemp, flax, sisal, and cotton. Carbon, glass, graphene, and Kevlar are examples of synthetic fibers used in AM composites.

Besides offering a high strength-to-weight ratio, fiber-reinforced polymer composites demonstrate high durability, stiffness, flexural strength, damping property, and resistance to corrosion, fire, impact, and wear. Natural fiber-reinforced polymers have the potential to move toward more environmentally friendly applications, given their biodegradability and low cost.

Carbon-fiber is employed when more stiffness is required, and carbon-reinforced polymer composites have many aerospace, automobile, sports, and other applications. When polymers combine with carbon fiber in AM applications, mechanical performance is significantly improved in tensile strength and flexural properties while also retaining dependence on build direction and AM process.

Graphene fibers are a newer high-performance fiber that shows higher tensile strength and more enhanced electrical conductivity than carbon fibers. These properties are well-suited for applications in lightweight conductive cables, knittable supercapacitors, solar cell textiles, and micromotors.

AM Notable 4.2 - Chicken Feathers – Not Just for Pillows Anymore!

One million chickens can produce up to 120 tons of chicken feather waste per year. As luck would have it, a chicken feather has good physical and mechanical properties and is ideal as an ingredient in epoxy resin composites to enhance tensile properties. Composites made with chicken feathers, which are 91% keratin protein, prove stronger than similar materials made with starch or soy proteins, making them commercially viable.

After pulverizing cleaned chicken feathers into fine dust, chemicals combine to join keratin molecules into long chains, known as *polymerization*. The result is a thermoplastic that uses no fossil fuels, is quickly heated and formed into shapes, and is highly biodegradable. As prices rise on petroleum products, bioalternatives are likely to gain popularity (Verma, Negi, and Singh, 2018).

Chicken feather fibers are integrated into MDF wood composite panels in factories experiencing a shortage of materials. Feather polymer composite materials could soon be used in cups, furniture, toys, helmets, automotive interiors, and a wide variety of other uses.

Metal Composites

Metal-composite filaments contain roughly 40–60% fine metal powder combined with PLA. The most common materials available as metal filament composites include copper, bronze, stainless steel, and iron. The metal-composite filament's main value is aesthetic, as they do not meet the performance of plastic filaments, nor do they possess the strength or durability of metal materials. Parts printed with the filament appear to be printed from metal and are heavier than plastic parts. Post-processing is required to generate the metallic-looking finish.

Metal matrix composites (MMCs) are a relatively new group of advanced composite materials that embed the excellent strength of ceramic reinforcements with metal or alloy. This combination helps provide strength and stiffness to enhance metal properties. Metals used include nickel, aluminum, and titanium. These composites have excellent physical and mechanical properties, making them ideal for aerospace, automotive, and structural applications (Larimian and Borkar, 2019).

Nonmetals

Ceramics

Ceramics represent one of the newer materials for AM. Ceramics are available in a liquid, powder, or solid format that can be processed by various AM systems. As a liquid or semi-liquid slurry, fine ceramic particles disperse as feedstock from an AM system. Powder-based ceramic materials contain loose ceramic particles that are processed using one of several AM powder bed systems. In solid form, sheets of ceramic material feed into AM sheet lamination systems.

Table 4.3 Composite Material Properties

Material	Application	Performance	Aesthetics	Geometry
Agilus Black	Flexible, rubber-like resin with superior tear resistance suitable for prototypes of rubber components such as seals, nonslip surfaces, etc.	**D/SH:** 1.14–1.15 g/cm³ Scale A: 30–35 **TS/TR:** 2.4–3.1 MPa4-7 Kg/cm **HDT:** 62°C @0.45 MPa 54°C@1.82 MPa **EB:** 220–270%	Matte finish	**SA:** 0.1–0.3 mm **LT:** 0.032 mm **MWT:** 1 mm **IE:** Yes
Alumide/PA-AF	Blend of aluminum and polymide powders produce nonporous, metallic-looking components; used in automotive industry, jig manufacturing, and models.	**D:** 1.36 ± 0.05 g/cm³ **TS:** 48 ± 3MPa **HDT:** 130°C **EB:** 3.5 ± 1%	Grainy surface with multiple finishes possible	**SA:** ±0.3% **LT:** 0.15 mm **MWT:** 1 mm, 0.3 mm possible **IE:** Yes
PA-GF	Polymide powder filled with glass particles with high thermal heat resistance; used for functional testing with high thermal loads and in demanding conditions.	**D:** 1.22 ± 0.03 g/cm³ **TS:** 51 ± 3MPa **HDT:** 110°C **EB:** 6 ± 3%	Grainy surface with multiple finishes possible	**SA:** ±0.3% **LT:** 0.12 mm **MWT:** 1 mm, 0.3 mm possible **IE:** Yes
Nylon 12 Carbon fiber	Nylon reinforced with chopped carbon fibers used for strong, stiff, lightweight parts and tools; can replace heavier metal parts.	**D:** 1.19 ± 0.03 g/cm³ **TS:** 83 ± 2MPa **HDT:** 160°C @66 psi **EB:** 2.4 ± 0.3%	Grainy surface with multiple finishes possible	**SA:** ±0.3% **LT:** 0.12 mm **MWT:** 1 mm, 0.3 mm possible **IE:** Yes

Source: Derived from Senvol (2019) and SpecialChem (2020).

Ceramics used in 3D printing processes can produce highly complex shapes with smooth and glossy surfaces. They are more durable than metal and plastic because they withstand pressure and very high temperatures without breaking or warping. Ceramics are also not prone to corrosion, like many metals, and do not wear like plastics. They are also available in a wide range of colors (Chen et al. 2019).

There are several challenges to using ceramics in AM. The material requires high temperatures to process and is not suitable for glazing or kilning. Since ceramics are fragile, they are unsuitable for printing objects with enclosed or interlocking parts and are not ideal for a part assembly process.

Ceramics are also used in conjunction with polymers and other materials, even seashells, to make composites with greater strength and less fragility.

Plaster

Plasters used in AM are a mix of dry powder and water that form a paste, which is the feedstock used for plaster-based AM. Plaster-based 3D printing uses inkjet heads like those found on home 2D inkjet printers and has been in use in AM for almost 30 years.

Plaster is ideal for some parts designed for relatively low-volume investment casting. AM processes used to 3D print a plaster master pattern create a disposable, cost-effective plaster mold to hold the molten metal. The plaster molds are quickly and inexpensively modified.

Plaster-based 3D printing is also an excellent approach for printing artistic projects in full color. No support structures are necessary for plaster-based AM, and it is relatively faster and less expensive than other materials and AM processes. The mechanical strength of plaster objects is lower, making plaster ideal for prototyping or plaster casting (Ayres et al. 2019).

AM Notable 4.3 - 3D Print a House in 24 Hours for $10k (or less)

ICON, the first company in America to be granted a construction permit for the purpose, built a 3D printed home in Austin, Texas, using 3D printing robotics, custom software, and advanced materials. While the prototype house cost $10,000 and build time was 24 hours, ICON says it can bring the cost down to about $4,000. ICON uses the Vulcan 3D printer to produce one-story buildings up to 2,000 square feet (Bendix, 2019).

Lavacrete, a proprietary Portland cement-based mix, is used by the Vulcan printer. The material, a carefully guarded secret, consists of easy-to-source raw materials and advanced additives. It has a compressive strength of 6,000 psi, which is stronger than most building materials. Lavacrete has a high thermal mass and handles extreme weather conditions to minimize natural disaster impacts. The wall system made with Lavacrete replaces traditional cladding, framing, and sheetrock to produce quality at a much lower cost (*Technology | ICON*, 2018).

3D printing just may be construction's digital revolution!

Sand

Sand is most frequently used in AM to 3D print molds and cores for casting objects from molten metals. Molds and cores are expensive to design and produce in traditional casting operations and have a significant lead time. Sand, applied with binder jetting, is used to 3D print molds and cores. Support structures are not necessary.

Figure 4.2 Sand 3D Printed Mold Package in Furan and Hot Hardening Phenol
Source: Courtesy of The ExOne Company.

The use of sand and AM in mold production has significantly reduced the lead-time and cost of mold-making. Plus, the AM process makes it easier to make improvements and design changes. The complexity of the object geometries is much higher with 3D printing versus the traditional casting process. Sand molds and cores can be printed on demand and used in the casting process. Figure 4.2 shows AM creation of a sand model.

Biomaterial and Organics

Biomaterials and organics are a high-growth area within AM. Biomaterials, which may be natural or synthetic, are those substances that have contact with biological systems. Organics, which are largely natural polymers or food-based substances, have recently been the focus of material science as new AM technologies are evolving to enhance their properties for use in a wide variety of applications (Whyte et al., 2019).

Bioprinting, 3D printing with biomaterials, has allowed medical researchers and manufacturers to create materials for research and patient-specific applications. Evolving AM medical applications make functional tissues to replace diseased or injured tissues. These applications rely on bioinks, a mixture of the required cell types and related materials in a hydrogel form, used in the bioprinting process (Gungor-Ozkerim et al., 2018).

Ceramics, hydrogels, metal pastes, silicones, and thermoplastics have been used, frequently alongside human tissues, since 2002. In response to the increasing demand for biomaterials, materials in different grades are now available, classified as technical-grade, research-grade, and medical-grade, with varying purity (EnvisionTEC, 2017). These materials, designed specifically for 3D-Bioplotter printers by EnvisionTEC, are briefly described below.

UV Silicone 60A MG is transparent and is biocompatible, bioinert, and nonbiodegradable. It is a UV light-cured liquid silicone rubber approved for short term use of fewer than 29 days in the human body. Wound dressings, microfluidics, biosensor housings, and prototyping are the primary applications of this material.

HT PCL 120K MG is a biodegradable, medical-grade thermoplastic polyester known as polycaprolactone (PCL). Processed at high temperatures, PCL excels in bone and cartilage regeneration applications, drug release, and hybrid scaffolds. It is appropriate for both short term and long term (over 29 days) use in the human body.

PCL 45 RG is a highly versatile PCL most frequently used for tissue engineering applications. It experiences almost no thermal degradation and is an ideal material for large, time-consuming parts.

Current AM research and development focuses on various other materials and composites, including optical-grade glass, high-temperature nylons, concrete composites, soft and flexible elastomers, fire, smoke, and toxicity compliant materials. Detailed information about proprietary materials offered by AM technology vendors is available via their websites. Chapter 10 contains additional sources of information about AM materials and guides to their selection.

Table 4.4 provides a summary of key AM materials and the AM technologies with which they are compatible. Each of the technologies will be explained and cross-referenced with materials in Chapter 5.

Table 4.4 Materials Used in AM

Material Category	Material Type	AM Technologies
Polymers	ABS (Acrylonitrile Butadiene Styrene)	Fused deposition
	ASA (Acrylonitrile Styrene Acrylate)	modeling (FDM)
	Nylon (Polyamide)	Stereolithography (SLA)
	PLA (Polylactic Acid)	Selective Laser
	PVA (Polyvinyl Alcohol)	Sintering (SLS)
	Polycarbonate	Material Jetting (MJ)
	Thermoplastics:	Multi Jet Fusion (MJF)
	High-performance polymers	Binder Jetting (BJ)
	Thermosets:	
	Thermoplastic polyurethane (TPU)	
Metals	Aluminum	Direct metal
	Aluminum Alloys	deposition (DED)
	Titanium	Direct metal laser
	Titanium Alloys	sintering (DMLS)
	Steel	Direct metal laser
	Precious metals: Gold and silver	melting (DMLM)
	Cobalt Chrome Alloys	Electron beam
	Nickel-based Alloys	melting (EBM)
		Laser metal
		deposition (LMD)
		Selective laser
		melting (SLM)
		Binder jetting (BJ)

Table 4.4 *(Continued)*

Composites	Thermoplastics with carbon, glass, or Kevlar fibers	Composite-based additive manufacturing (CBAM)
	Metal and diamond	Powder bed fusion (PDF)
	Graphene nano-composites	Material extrusion (ME)
Ceramics	Alumina	NanoParticle Jetting (NPJ)
	Zirconia	
	Silica glass	
	Porcelain	
	Silicon-Carbide	
Other Non-Metals	Sand	Binder Jetting (BJ)
	Wax	Material jetting (MJ)
	Concrete	
BioMaterials	Rice paste	Material Extrusion (ME)
	Foods	
	Bone cells	
	STEM cells	
	Human tissues	
	Hydrogels	
	Aginate	
	Chitosan	
	Gelatin	
	Cllagen	
	Fibrin	

Source: Original, Kamara and Faggiani, 2020.

Case Conclusion: Material Selection and Testing

After learning more about the available AM materials, the GWM AM Pilot Project team identified stainless steel as the best candidate for the camp stove element. It is lightweight, has good thermal properties, is relatively easy to source, and is compatible with several AM processes and technologies. They also discovered that 3D printed stainless steel is certified as a gas conduit as needed in their application.

The selection of material helped narrow down the number of service bureaus they could approach, since not all of them had metal AM capabilities. Their next step is approaching a service bureau with metal capabilities to discuss creating a functional prototype that they could further test.

Unfortunately for GWM, the closest service bureau to their location did not have metal 3D printing capability. The team discussed the possibility of holding virtual meetings with representatives of the closest service bureau with metal capabilities, a two-hour drive away. They also considered taking a day trip to the service bureau. Given the variety of questions that needed answers and the fact the team had no experience in metal additive manufacturing, they decided to make the trip for an in-house meeting, for the following reasons:

- **Evaluate benchmark parts**. Even without producing their specific parts, a physical visit would allow the team to see other benchmark or sample parts printed by the service bureau. They would be able to see unfinished parts, with and without supports, to understand the process further and ask meaningful questions.

- **Customer relations**. Should they identify similar components produced for other customers of the service bureau, they could use the other service bureau customers as a resource to avoid costly mistakes. A noncompeting customer willing to share their pain points could benefit the team's effort in improving their AM knowledge and managing expectations.

- **Validate expertise.** Not all service bureaus are equal. The setup of the bureau's facility, systems, and post-processing equipment availability could be invaluable for assessing their capacity for potential production work in the short-term.

REFERENCES

Technology | ICON (no date). www.iconbuild.com/technology (accessed: 9 November 2020).

AMFG (2020). The top 10 3D printing trends to expect in 2020. AMFG. https://amfg.ai/2020/01/07/top-10-3d-printing-trends-in-2020/ (accessed: 9 November 2020).

Ayres, T., Sama, S., Joshi, S. et al. (2019). Influence of resin infiltrants on mechanical and thermal performance in plaster binder jetting additive manufacturing. *Additive Manufacturing,* 30, doi: 10.1016/j.jeurceramsoc.2018.11.013.

Bendix, A. (2019) 3D homes that take 24 hours and less than $4,000 to print. *Business Insider* (March 12). www.businessinsider.com/3d-homes-that-take-24-hours-and-less-than-4000-to-print-2018-9 (accessed: 9 November 2020).

Chen, Z, Ziyong, L., and Chengbo, L. et al. (2019). 3D Printing of ceramics: A review. *Journal of the European Ceramic Society,* 39 (4), pp. 661–687. doi: 10.1016/j.jeurceramsoc.2018.11.013.

EnvisionTEC (2017). Bioprinting materials: From research to clinical use. 3D printing materials. https://envisiontec.com/3d-printing-materials/bioprinting/ (accessed 14 August 2020).

Goldschmidt, B. (2020) 3D printer material cost: The real cost. All3DP. https://all3dp.com/2/3d-printer-material-cost-the-real-cost-of-3d-printing-materials/ (accessed: 29 October 2020).

Gungor-Ozkerim, P. S. et al. (2018). Bioinks for 3D bioprinting: an overview. *Biomaterials Science,* 6 (5), pp. 915–946. doi: 10.1039/c7bm00765e.

Larimian T., Borkar T. (2019) Additive manufacturing of in situ metal matrix composites. In: AlMangour B. (eds) *Additive Manufacturing of Emerging Materials*. Springer, Cham. https://doi.org/10.1007/978-3-319-91713-9_1

Lee, J.-Y., An, J., and Chua, C. K. (2017). Fundamentals and applications of 3D printing for novel materials. *Applied Materials Today, 7*, 120–133. doi:10.1016/j.apmt.2017.02.004

Materialise (2020). ABS. www.materialise.com/en/manufacturing/materials (accessed on 12 August 2020).

Metal AM (2020). Metal additive manufacturing/3D printing: An introduction. www.metal-am.com/introduction-to-metal-additive-manufacturing-and-3d-printing/ (accessed 14 August, 2020).

Rajak, D. K., Pagar, D. D., Menezes, P. L., and Linul, E. (2019). Fiber-reinforced polymer composites: manufacturing, properties, and applications. *Polymers,* 11(10), 1667. https://doi.org/10.3390/polym11101667

Relativity Space (2020). Stargate. www.relativityspace.com/stargate (accessed: 9 November 2020).

Senvol (2019). Senvol database: Industrial additive manufacturing machine and materials. materials search. http://senvol.com/material-search/ (accessed 06 August 2020).

Sher, D. (2017). How composite materials are changing the world of additive manufacturing (again). 3D Printing Media Network. www.3dprintingmedia.network/composite-materials-changing-world-additive-manufacturing/ (accessed 12 August 2020).

SpecialChem (2020). Omnexus 3D printing/additive manufacturing using polymers – Complete guide. https://omnexus.specialchem.com/selection-guide/3d-printing-and-additive-manufacturing-polymers-and-processes#MainPolymers (accessed 09 August 2020).

SpecialChem (2020). Omnxus Universal Selector by SpecialChem. https://omnexus.specialchem.com/selectors (accessed 09 August 2020).

Stratasys Direct (2020). Choosing the right material for your application. White Paper. www.stratasysdirect.com/resources/white-papers/how-to-choose-3d-printing-materials (accessed 16 July 2020).

Verma, A., Negi, P., and Singh, V. K. (2018a). Experimental investigation of chicken feather fiber and crumb rubber reformed epoxy resin hybrid composite: mechanical and microstructural characterization. *Journal of the Mechanical Behavior of Materials,* 27 (3–4). doi: 10.1515/jmbm-2018-0014.

Verma, A., Negi, P. and Singh, V. K. (2018b). Experimental investigation of chicken feather fiber and crumb rubber reformed epoxy resin hybrid composite: mechanical and microstructural characterization. *Journal of the Mechanical Behavior of Materials,* 27 (3–4). doi: 10.1515/jmbm-2018-0014.

Whyte, D., Rajkhowa, R., Allardyce, B. and Kouzani, A. (2019). A review on the challenges of 3D printing of organic powders. *Bioprinting,* 16, e00057,ISSN 2405-8866, https://doi.org/10.1016/j.bprint.2019.e00057.

Chapter 5

Which Additive Manufacturing Technology and Process Are Right for My Solution?

Case Introduction: Selecting an AM Process and Technology

The GWM AM Pilot Project team has selected stainless steel for their camp stove heating element. It is investigating an AM process and technology that can help fulfill their design intention. They are meeting with Doug Blackstone from an AM service bureau to discuss their options. Doug has had the chance to review the part design and the full requirements for the part and camp stove. Following is a snippet of the conversation:

Fundamentals of Additive Manfacturing for the Practitioner, First Edition. Sheku Kamara and Kathy S. Faggiani © 2021 John Wiley & Sons, Inc. Published 2021 by John Wiley & Sons, Inc.

DOUG:	You've all done your homework. That much is clear. And congratulations on the smart design and your first foray into AM!
BOB:	Thanks, Doug. We appreciate your willingness to meet with us. As we're new to AM, we need to understand our AM process and technology options to produce this part to meet all requirements.
DOUG:	Your focus on stainless steel is right on target. Only two AM technologies do not direct print metal, so you'll have some options. AM processes will leverage the thermal properties to ensure the final part is lightweight and ideal for your application. You do have a wide choice of AM processes and technologies for metal, but I can tell you upfront that the following will most likely meet your needs:

- Binder jetting

- Directed energy deposition

- Material extrusion

- Powder bed fusion

- Sheet lamination

| | I have an idea, but since you're all new to this, let's take the time to go over each process and discuss the advantages and disadvantages. Then we can get into the costs for your functional prototype. |
| ROXIE: | Sounds like a plan, Doug! (The team members nod in agreement.) |

INTRODUCTION

The formal definition of additive manufacturing (AM) is: process of joining materials to make parts from 3D model data, usually layer upon layer, as opposed to subtractive manufacturing and formative manufacturing methodologies (ISO/ASTM 52900:2015(en), Section 2.1.2).

At present, AM processes divide into seven widely recognized categories: binder jetting, directed energy deposition, material extrusion, material jetting, powder bed fusion, sheet lamination, and vat polymerization. An eighth category, hybrid processing, has recently evolved and is rapidly becoming mainstream. Hybrid processes utilize both additive and subtractive techniques within the same machine to produce the desired part.

Several variations exist within AM processes. For metal AM, the process may be direct or indirect. In direct metal processing, metal parts or components are produced directly from an AM system. In indirect metal processing, AM is used to create a sacrificial part, mold, or green part that needs to be sintered or cured to achieve the desired properties. The technologies used within some processes may also vary as existing AM technologies continue to evolve, and new technologies develop for existing operations. Figure 5.1 illustrates current technology systems provided by selected vendors and will be referred to throughout this chapter.

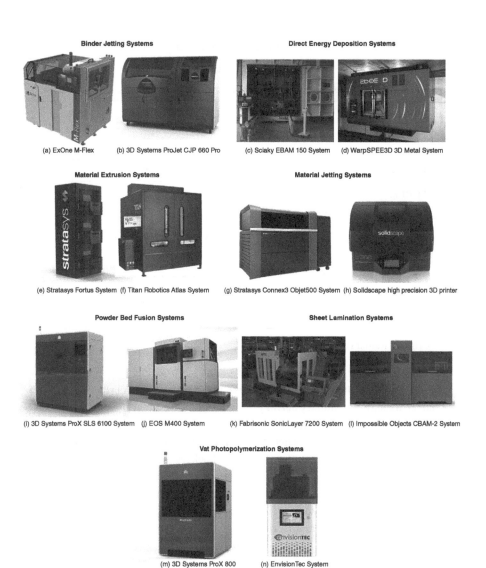

Binder Jetting Systems

(a) ExOne M-Flex   (b) 3D Systems ProJet CJP 660 Pro

Direct Energy Deposition Systems

(c) Sciaky EBAM 150 System   (d) WarpSPEE3D 3D Metal System

Material Extrusion Systems

(e) Stratasys Fortus System   (f) Titan Robotics Atlas System

Material Jetting Systems

(g) Stratasys Connex3 Objet500 System   (h) Solidscape high precision 3D printer

Powder Bed Fusion Systems

(i) 3D Systems ProX SLS 6100 System   (j) EOS M400 System

Sheet Lamination Systems

(k) Fabrisonic SonicLayer 7200 System   (l) Impossible Objects CBAM-2 System

Vat Photopolymerization Systems

(m) 3D Systems ProX 800   (n) EnvisionTec System

Figure 5.1 AM Vendor Systems

Source: (a) Courtesy of The ExOne Company; (b) Courtesy of 3D Systems, Inc.; (c) Courtesy of Sciaky; (d) Courtesy of SPEE3D, (e) Photo Credit: Stratasys; (f) Courtesy of Titan Robotics; (g) Photo Credit: Stratasys; (h) P20, Solidscape high precision 3D printer; (i) Courtesy of 3D Systems, Inc.; (j) Image courtesy of EOS; (k) Courtesy of Fabrisonic LLC; (l) Courtesy of Impossible Objects; (m) Courtesy of 3D Systems, Inc.; (n) Photo used courtesy of EnvisionTEC.

Table 5.1 Summary of AM Technology and Associated Materials

Technology/ Materials	Polymers	Metals	Composites	Non- metals	Biomaterials
Binder jetting	X	X	X	X	X
Directed energy deposition		X			
Material extrusion	X	X	X	X	X
Material jetting	X	X	X		X
Powder bed fusion	X	X	X	X	X
Sheet metal lamination		X	X		
Vat photopolymerization	X			X	X
Hybrid		X	X	X	

Source: Original, Kamara and Faggiani, 2020.

This chapter presents all eight of the AM process categories and their process and technology variations. It also includes a discussion of materials appropriate for each process and technology. A summary of AM processes applicable to the material categories presented in Chapter 4 appears in Table 5.1.

AM technology and process, as noted in the preceding chapter, impact materials and mechanical properties of the resulting parts. This book assists the reader in achieving the optimal AM design solution to any manufacturing challenge. In reality, the final design relies on the process. Some modifications to the optimal solution may be necessary to achieve the required geometry and performance for a given application. Plans may be needed to achieve the desired aesthetics through post-processing.

BINDER JETTING

Binder jetting systems utilize printheads to stream a fusing agent onto a bed of powder, layer by layer, similar to a 2D printer. A liquid binding agent is typically sprayed through customized or HP printheads, printing cross-sections of the part (or parts) layer by layer. A new layer of powder spreads across the platform as the process repeats itself. Binder jetting is considered an indirect metal process since it generates molds, cores, or sacrificial parts used in castings. Additionally, metal powder is fused with the binding agent and subsequently sintered to full density. Figure 5.1(a) and (b) show two popular binder jetting systems.

Figure 5.2 shows a graphic representation and machine image of the binder jetting process. Powdered material, either ceramic, metal, polymer, or sand, fill the powder chamber (A) of the machine. The powder chamber indexes vertically to allow the roller (B) to spread powder onto the build platform (D). When the powder particle size falls below 13 microns, ultrasonic technology and dual rollers spread the powder smoothly and uniformly. After applying a few layers, the customized or HP printhead (C) moves across the build platform, jetting the part's cross-section as defined in the digital file onto the powder. Because the printhead only wets and binds the cross-section using the liquid binding agent, support structures are not required since the loose unbonded powder supports the new sections of subsequent layers.

A key advantage of binder jetting is throughput. The print time per layer is generally constant, irrespective of the cross-section or number of parts produced. The build time is directly related to the number of layers printed per the height of the part(s) based on the Z-orientation. For example, Voxeljet's VX4000 sand printer has a build volume

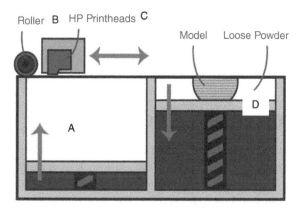

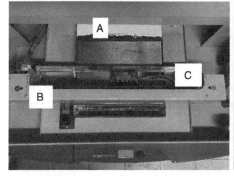
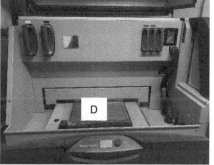

Figure 5.2 Binder Jetting Process Flow
Source: Illustration reprinted with permission from Tooling U-SME®; photos courtesy of MSOE.

of 4000 mm × 2000 mm × 1000 mm (13 ft × 6½ ft × 3¼ ft) with a vertical speed of 15.4 mm/h (0.6 inches/h) (Voxeljet, 2015). The size supports printing multiple parts with different heights at the same time, thereby reducing throughput time.

One of the limitations of the process is the low strength of the green parts produced. Green parts require further processing after printing, usually sintering for ceramic and metal powders or curing for polymer and sand powders. The lower green strength presents a challenge in removing unused powder from within the part with some materials. The build chamber temperature is generally at room temperature, and the unused material does not experience a phase change; hence 100% of the unused material is recyclable.

Example: Creating a Futuristic Aircraft Door with Binder Jetting

Sogeclair, a French aviation supplier, produced investment casting patterns for an aircraft door. They printed the polymethyl methacrylate (PMMA) models on Voxeljet's VX1000 system. The design was impossible for traditional methods to produce the patterns. With a build area of 1,000 × 600 × 500 mm and standard layer thicknesses of 150 μm (0.006 in) with a print resolution of 600 dpi, the VX1000 binder jetting machine produced the parts with increased precision and resolution. The PMMA parts were then coated with wax and burnt out at 700 °C after coating with the ceramic shell (Voxeljet 2017, Voxeljet Services, 2020).

DIRECTED ENERGY DEPOSITION

Directed energy deposition (DED) utilizes focused energy to fuse layers of metal material supplied as powder or wire feedstock. The energy comes from a laser or electron beam, gas tungsten or plasma arc, or compressed air at supersonic speed. Classified as a direct metal process, parts from the process do not require sintering or curing to achieve desired mechanical properties. A valuable attribute of DED is its ability to deposit material on existing structures for repair or strengthening. Figures 5.1(c) and (d) display two DED systems.

Figure 5.3 depicts the graphic representation and machine image of a DED process. The material feeds through a nozzle (A). In this illustration, the material is in wire filament form. As the wire extrudes, thermal energy from the electron beam (B) melts the metal onto the build platform (C). Another nozzle houses a second or similar material to increase throughput.

One of the key advantages of directed energy deposition is throughput for direct metal AM systems. For example, Sciaky's EBAM systems can deposit up to 11.34 kg (25 lbs.) of metal per hour (Sciaky, 2020).

Some of the limitations of the DED process are accuracy and surface finish. The technique excels at producing parts that are near-net shaped and must be post-machined for accuracy and surface finish. Hybrid systems that combine additive and subtractive technologies within the same platform make it easier to move from part printing to part finishing, as discussed in the section on hybrid technologies.

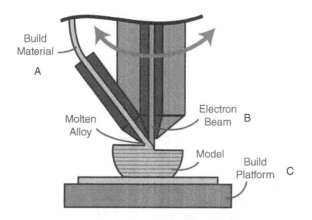

Figure 5.3 DED Process Flow

Source: Illustration reprinted with permission from Tooling U-SME®; photos courtesy of Sciaky.

Example: Giant Satellite Fuel Tank

Lockheed Martin needed a 46-inch (1.16-meter) diameter titanium dome for satellite fuel tanks to complete final rounds of quality testing. The fuel tank was part of a multiyear development program to create giant, high-pressure tanks that carry fuel aboard satellites. The tanks must be strong and lightweight to hold up to the rigors of launch and missions in the vacuum of space. Titanium was the perfect material, but acquiring the raw material and forging the tank takes more than a year with traditional manufacturing techniques.

By moving to 3D printing with DED instead of traditional manufacturing, where tank production required 18–21 months and costs $330–375k per tank, Lockheed saved considerable time and cost. Using Sciaky's directed energy deposition system, it took 5–6 months at a lower cost of $170k per tank. With five tanks needed per vehicle, the total cost savings for Lockheed Martin was 55%, and time reduction was 66%. The giant titanium fuel tanks were the largest 3D printed space parts made at the time (Lockheed Martin, 2018).

AM Notable 5.1

Walrus Dentistry with PBF

Canada houses four walruses at the Aquarium du Quebec as a conservation effort. Walruses use their tusks to pull themselves from the water, and rather than encountering ice or soil in their new habitat, their tusks grind against concrete, metal, or other materials. The risk of fracturing or cracking the mammals' tusks is high. Rather than removing the tusks, veterinary medical experts recommended made-to-measure crowns or caps for the tusks.

Two Canadian professors collaborated with veterinarians to design and create metal walrus tusk crowns. They chose cobalt chrome, a biocompatible alloy that would not oxidize in saltwater and was hard enough to stand up to repeated compression and rubbing. Direct metal laser sintering, a type of directed energy deposition process, was selected to build the tusks. A 3D scanner produced a digital image of the walrus tusks as the input to the printing process.

The scan was modified to provide a design with perfect clearance between tusk and crown for the bonding glue to adhere the metal to the tusk (Crevier, C., 2020). The metal tusks were custom-printed and affixed without anesthesia after animal trainers taught the walruses to remain immobile for the five minutes needed for installation.

Numerous other examples of eco-friendly AM applications in wildlife conservation exist. A conversation organization stymied the theft and trade of sea turtle eggs in Costa Rica by 3D printing eggs containing trackers, leaving them in nests, and using them to uncover illegal trade routes. A 3D-printed Slothbot, created at the Georgia Institute of Technology, monitors the temperature, weather, and carbon dioxide levels at the Atlanta Botanical Gardens to help preserve the local marine park. More unique applications of AM are on the horizon (Hanaphy, 2020).

MATERIAL EXTRUSION

Material extrusion utilizes polymer or polymer-metal composite packaged in a filament or rod extruded through a heated nozzle that softens the material. The nozzle moves over the build platform, depositing a thin bead of extruded material to form each layer. The material hardens as it extrudes and bonds to the layer below. Most industrial systems include a specific nozzle for the model material and one for the support material. Others have multiple nozzles to incorporate colors and different materials during the build. Most low-cost kits and systems use a single nozzle for both part and support generation. Figures 5.1 (e) and (f) display two material extrusion systems.

Figure 5.4 includes a graphic representation and machine image of material extrusion. Material feeds from the feeder roll (D) to the heated extruder (A and B), where it heats to a liquid or semi-liquid state. In this illustration, model material deposits on the build platform (C). The model geometry determines how the model and support extruder will extrude material alternatively on the same layer. For systems with a single extruder, the model material is used as support material as well.

Relatively lower equipment costs, the widespread use of the technology, and availability of materials are advantages of the material extrusion process. Material extrusion has the largest selection of available polymer materials for AM, and the list now includes ceramics and metals.

One of the main limitations of material extrusion is its elongation-at-break in the build direction. In material extrusion, the layers are melded on top of each other and do not overlap as do laser-based processes. Hence the difference in elongation-at-break between the build direction and the build plane could be over 20% (Stratasys, 2020c). For this reason, understanding the process and design requirements for the part or component will determine the part orientation on the build platform.

Material Extrusion

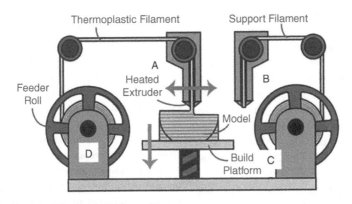

Figure 5.4 Material Extrusion Process Flow
Source: Illustration reprinted with permission from Tooling U-SME®; photos courtesy of MSOE.

Example: Precision Parts for Production with Material Extrusion

ZEISS Optical Components is a manufacturer of microscopes, optical sensors, and other high-precision items for industrial measurement and quality control applications. ZEISS was very concerned with accuracy and reliability. Using Ultimaker 3D printers to produce custom adapter plates, which ensure that light travels precisely in the required device's sensors, ZEISS generated unique adapter plates for microscopes in serial production. ZEISS currently 3D prints a unique adapter plate for every microscope in serial production. 3D printed jigs and fixtures, customized to calibrate each device, are also produced. The cost per part dropped from €300 to €20 for each adapter plate while maintaining repeatable accuracy (Ultimaker, 2020).

AM Notable 5.2

Edible AM – Direct and Indirect Processes

Remember the Play-Doh Fun Factory? While it didn't produce edibles (you've probably tried telling a two-year-old that), it is a great example of the direct process of material extrusion applied to food to create edibles. Many foods are suitable for extrusion. With chocolate's physical characteristics, it is the most commonly extruded food; also used are cream cheese and mashed potatoes. By combining different types of foods, Foodini created an extruded AM pizza.

Binder jetting, another direct printing process, can produce complex, detailed sugar sculptures. 3D Systems created printers that replaced metal or polymer powders with sugar crystals, using a mix of water and alcohol as the binding agent. Since binder jetting can create detailed logos and photos, it presents branding and product differentiation opportunities for confectioners' products.

The indirect approach to AM-enabled food production is mold printing. Vat photopolymerization processes, such as stereolithography, are used to print food-safe, FDA-approved materials in shapes. These reusable molds shape food substances. For example, flavored gelatins begin in solid form and cannot be extruded; instead, the printing of molds to shape gelatin into custom designs offers product customization options (Porter, Phipps, Szepkouski, et al., 2015).

By changing materials and modifying AM technologies, AM processes can readily be adapted to produce edibles. Beyond the support AM provides the food industry for product differentiation, product customization, and direct-to-consumer relationships, imagine how AM-enabled food products could contribute to our efforts to explore space!

MATERIAL JETTING

Material jetting utilizes customized printheads to deposit tiny droplets of photopolymer, ceramic, or metal nanoparticles suspended in liquid onto a build platform. A photopolymer layer immediately cures by exposure to UV light. Layers of ceramic or metal particles are heated to evaporate the liquid surrounding the particles. As the printhead traverses the platform, part and support materials jet onto the successive layers. The support material, usually water-soluble, is easily removed from the model. Although similar to binder jetting, the time per layer is not constant. The material is only applied where the parts are and not across the entire platform. Recoating the whole platform with powder is not necessary. Figures 5.1 (g) and (h) show two commercial material jetting systems.

Figure 5.5 includes a graphic representation and machine image for a polymer-based material jetting machine. Droplets of material are jet from the printhead (A) onto the platform (C). The UV light (B) attached to the printhead immediately floods light across the platform, curing the droplets and converting them to a solid. The process repeats for each layer.

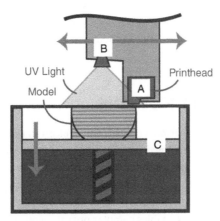

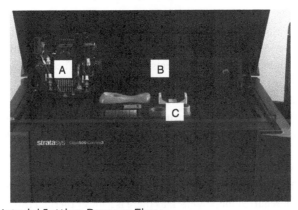

Figure 5.5 Material Jetting Process Flow
Source: Illustration reprinted with permission from Tooling U-SME®; photo courtesy of Stratasys.

One of the advantages of material jetting is the high resolution of parts. Material jetting has the highest resolution of AM technologies, especially on the z-axis of the build platform. With its high resolution, multiple material capabilities, and color options, material jetting produces near photorealistic parts.

One of the limitations of material jetting is material cost. The process utilizes part and support material in similar proportions. However, some part geometries require more support material than part material when printing stability requires complete part enclosure.

Example: Printing Material Packaging with Material Jetting

PPI-Time Zero manufactures expensive motors shipped directly to clients in the defense, aerospace, and security industries. To protect the motors from damage in transit and keep them watertight for cleaning, PPI uses a 3D Systems ProJet® MJP 2500 Plus to produce a motor cover to seal and protect motors from contamination during cleaning and transport. The material used was Visijet Proflex, a durable and pliable substance easily removed by the client and returned to PPI for reuse (3D Systems, 2019).

POWDER BED FUSION

Powder bed fusion utilizes thermal energy from an electron beam, laser, or infrared heating to fuse, sinter, or melt powder layer by layer. The range of powdered material is diverse as it ranges from polystyrene to metal to ceramics. Though the process is similar for polymers and metals, no support structures are necessary for polymers though they are required for metals to hold the parts on the build platform. The elimination of polymer support structures enables complex organic geometries with minimal postprocessing except for removing unused powder. Figures 5.1 (i) and (j) display two powder bed fusion systems.

As shown in Figure 5.6, graphic representation and machine image, the process works by the roller (A) spreading a fine layer of powder on the build platform (C). The heat source, a laser in this illustration, heats the material to fuse it to reach the successive layer.

For metal-based powders, support structures or components are built directly on the build platform (C). Internal stresses during the build process make this necessary since the build chamber is at room temperature, and thermal energy from an electron beam or laser is solely responsible for sintering or melting the material. For polymer-based powders, the polymer in the build platform is heated to about 12 °C below its melting point, and the thermal source provides the additional heat for sintering or melting. Support structures are not required for polymers, as there is a lower thermal threshold compared to metals.

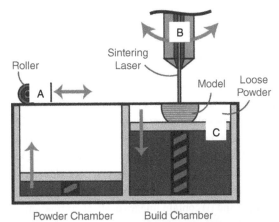

Powder Bed Fusion

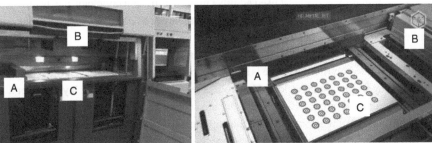

Figure 5.6 Powder Bed Fusion Process Flow

Source: Illustration reprinted with permission from Tooling U-SME®; photo on right courtesy of HP, photo on left courtesy of MSOE.

One of the advantages of powder bed fusion is its isotropic properties. Powder-based fusion parts tend to be stronger in the z-axis build plane than the other axis due to the extended thermal exposure. However, for polymers, a higher tensile strength produces an elongation at break that is slightly lower in the build direction.

Some of the limitations of powder bed fusion are equipment and facilities costs. In general, powder bed fusion technologies require additional post-processing equipment and safety requirements for powder handling and post-processing. Additionally, handling of fine powder particles in either polymer or metal form adds a potential safety hazard.

Example: Medical Research Device Creation with Powder Bed Fusion

A research study investigating motor control and brain activity in persons recovering from stroke needed a medical device compatible with functional magnetic resonance imaging (fMRI) technology. The handbike frame and shaft with a flange were designed in one piece to simulate a flange bearing while preventing the handle from disengaging during operations. A selective laser sintering (SLS) machine using Polyamide 12, low-water-absorbing nylon with good tensile and impact strength, printed the fMRI-compatible hand bike.

SHEET LAMINATION

Sheet lamination utilizes sheets of paper, long-fiber fabrics of carbon or glass, or metal foil connected with glue, thermoplastic matrix materials, or ultrasonic welding, respectively. Cross-sections cut with a knife or laser for paper, or a cutter on a three-axis mill for metals, form the part shape. For composite-based additive manufacturing (CBAM), the sheets of carbon and glass layers are stacked together, then heated to the melting point of the thermoplastic matrix used to fuse the layers. A bead blaster or chemical process removes uncoated fibers. Color parts that have a wood-like texture are printable using paper. Figures 5.1 (k) and (l) contain images of sheet lamination systems.

Sheet lamination is the least used additive manufacturing technology, partly due to limited application. Paper parts from these systems are used for design verification, marketing, and educational models.

Figure 5.7 provides a graphic representation and machine image of sheet lamination. Metal foil aligned on the base plate is subject to localized welding between the foil and base plate by an ultrasonic welder. The process repeats with subsequent layers. After a few layers, an integrated CNC milling machine cuts a profile of the component. The process continues until the completion of the entire part. Ultrasonic additive manufacturing (UAM) allows for multi-material applications as different alloys can be welded at various times along the build plane.

Some of the advantages of sheet lamination are the ease of material handling and scalability. For the UAM process, the hybrid process enables designers to create complex internal geometry, embed sensors and electronics, and produce new material combinations (Fabrisonic, 2020b).

The limitation of the technology is the additional steps needed to produce the final end-use parts. These extra steps will increase lead-time and cost for potential applications.

Example: Creating Thermal Management Devices with Ultrasonic Additive Manufacturing (UAM)

NASA is known for developing robust interstellar devices. However, for decades the thermal management systems on NASA's satellites and rovers have been constructed of bent metal tubes glued to the outside of a vehicle's chassis. Over several years, Fabrisonic worked with NASA to qualify UAM as the next technology for printing interstellar thermal management devices. The work resulted in a Fabrisonic heat exchanger that passed stringent NASA qualifications, including vibration, thermal, hermeticity, and burst tests.

Sheet Lamination

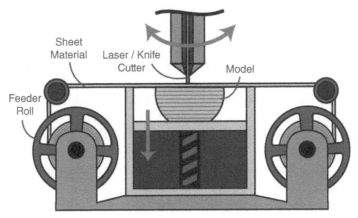

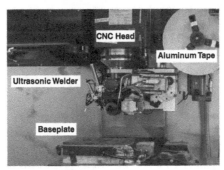

Figure 5.7 Sheet Lamination Process Flow
Source: Illustration reprinted with permission from Tooling U-SME®; photo courtesy of Fabrisonic, LLC.

VAT PHOTOPOLYMERIZATION

Vat photopolymerization utilizes photopolymers and an ultraviolet light source to selectively cure, change resin properties from liquid to solid, and create each cross-section of the part. Systems are either downward facing or upward facing, depending on the orientation of the build. The photopolymer is stored in a vat for downward-facing systems, as parts are built on a perforated platform allowing the resin to flow through. Upward-facing systems generally utilize digital light processing (DLP) to cure the entire layer simultaneously, increasing throughput. Support structures are required to attach the part to the build platform and are used for leveling should there be any issues with the build platform flatness. Figures 5.1 (m) and (n) illustrate vat polymerization systems.

Figure 5.8 provides a graphic representation and machine image of vat photopolymerization. Photopolymer resin (C) fills a vat to a specific predetermined height. The perforated build platform (A) is positioned slightly above the resin level, allowing the cured resin to adhere to the build plate. Support structures allow for easy removal of the parts. Generally, after 10 mm (0.4 in.) of build height, the recoating blade (B) moves across the resin level to provide a smooth surface before the UV source cures the resin. The process repeats itself layer by layer until the completion of the final part.

Some of the advantages of vat photopolymerization are its accuracy and surface finish. With different viscosities and mechanical properties of photopolymer resins, the photoinitiation occurs with minimal energy. The parts are easier to post-process for a superior finish.

Support and handling of additional equipment and chemicals to process the end-use part are limitations of vat photopolymerization. Just-built parts are considered green, with surfaces cleaned to remove uncured resins. During this period, safety precautions are in place to handle the parts before they fully cure in a post-cure apparatus.

Vat Photopolymerization

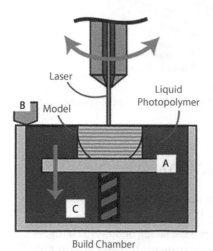
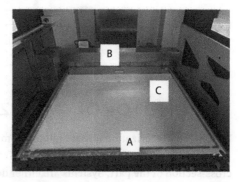

Figure 5.8 Vat Photopolymerization Process Flow
Source: Illustration reprinted with permission from Tooling U-SME®; photo courtesy of MSOE.

Example: 3D Printing of Hearing Aids with Vat Photopolymerization

AM is now the dominant manufacturing method for hearing aids and has significantly reduced the number of steps required to make a hearing aid. Traditional methods required an impression of the recipient's ear, from which a sculptor produces a positive cast of the ear in which to pour a plastic modeling material. Once complete, the whole project is cured in ultraviolet light to get a plastic piece. The plastic piece is drilled for vents and sound holes or is subject to additional manual processing to achieve the final shape.

Producing a hearing aid using AM significantly reduces the number of steps required and eliminates much manual processing. First, a silicone cast or impression of someone's ear is captured as a three-dimensional image by a digital scanner. The scan is then converted by CAD software specific to the hearing aid industry into an image file that a 3D printer can read. This three-dimensional image allows the user to alter the file to create the final product shape (EnvisionTec, 2020b).

AM Notable 5.3

Bioprinting Human Organs

The need for organ transplants has been driving ongoing bioengineering research efforts into bioprinting human organs. A major barrier to creating functional tissue replacements has been the difficulty in printing sophisticated vascular networks – airways and blood vessels – that supply nutrients to tissues. The development of a new bioprinting technology that addresses the need for multivascularization takes us one step closer to easing the shortage of transplant organs and saving lives.

The new open-source bioprinting technology is called SLATE, short for the stereolithography apparatus for tissue engineering. SLATE prints layers of a liquid pre-hydrogel solution that solidify when exposed to blue light. A digital light processing projector (DLP) shines light from below to display 2D slices of the structure at high resolution with pixel slices of 10–50 microns. As each layer solidifies, an overhead arm raises the growing 3D gel just enough to expose liquid to the next projector image. Photoabsorbers sensitive to blue light, added to the solution, help produce a very fine layer. Soft, water-based, biocompatible gels with complex internal architectures are produced in minutes to support the vascularization needed for functional tissues. Tests show the tissues are sturdy enough to support blood and oxygen flow and that red blood cells could take up oxygen as they flowed through a network of blood vessels (Rice University, 2019).

The hydrogels used in the bioprinting process can incorporate patients' cells, thus significantly reducing organ rejection and improving the safety of the transplant process. More challenges on the path to printing functioning human organs exist, though researchers believe the possibility will become a reality within the next 20 years.

HYBRID SYSTEMS

Hybrid systems are additive manufacturing systems that use both additive and subtractive methods within the same technology. These systems typically produce metal parts. Currently, hybrid systems exist in directed energy deposition, powder bed fusion, and sheet lamination technologies. CNC machining is used as a subtractive process in many metal hybrid systems to ensure part dimensional accuracy.

The DMG Mori's LASERTEC 65 3D hybrid system provides laser-powered direct material deposition and incorporates a milling machine to produce parts in finish quality (DMG Mori, 2020). Mazak's INTEGREX i-400 AM hybrid system uses fiber laser heat and multiple nozzles to produce shapes. It can join different types of metals to repair worn or damaged sections and deposit new material (Mazak, 2020). Following the deposition process, the near-net-shape part is finished with high-precision machining to achieve the desired tolerances. Parts are finished in one process within a hybrid system to increase throughput and produce increased complexity with an enhanced surface finish.

Matsuura's LUMEX Avance-60 hybrid system is an example of a powder bed fusion hybrid (Marsuura, 2020). The process machines every tenth layer as the parts are 3D printed. The periodic machining between layers improves the accuracy and surface finish of the part compared to a standard powder bed fusion system. The process can produce deep channels without secondary processes, like the use of an electrical discharging machine (EDM).

Case Conclusion

SELECTING AND JUSTIFYING AN AM TECHNOLOGY/PROCESS

The GWM AM pilot team initially identified five processes suitable for working with their camp stove fuel component and their selected metal. The processes are:

- Binder jetting
- Directed energy deposition
- Material extrusion
- Powder bed fusion
- Sheet lamination

The team assigned each member a process to investigate in further detail, and then the members met to discuss the findings with Doug from the AM service bureau. Part of the conversation follows:

FRANK: From a cost perspective, I sure hope we can avoid binder jetting and material extrusion. It sounds like both of those processes will require additional post-processing, and all I see are dollar signs adding up.

ROXIE:	You mean the post-build furnace cycle? Wouldn't that be done in the same machine, or is that process handled in a different device?
DOUG:	Hybrid is usually an AM process combined with a traditional subtractive process. Although parts are machined during the build process, the level of complexity is limited compared to powder bed fusion.
PETE:	It doesn't look like material extrusion will really work to get what we want for the design and function we need for the fuel component. Some complexity was added in the redesign, and the tolerance requirements could be affected in the furnace cycle as the material shrinks by a few percentage points. I don't think binder jetting, directed energy deposition, and material extrusion could easily meet the requirements. However, I was impressed that support structures could easily dissolve before entering the furnace cycle in material extrusion and are not required at all in binder jetting. What do you think about powder bed fusion?
BOB:	I agree with Pete. I think powder bed fusion is the direction to go. From my understanding of the technologies, it works with stainless steel powder. It will allow us to meet our tolerances and level of detail, plus it will create a part that passes muster with our regulatory requirements for gas conduits. Like that, Roxie?
ROXIE:	You bet I do!
DOUG:	So do I! That would be my suggestion as well. We've got a powder bed fusion system at the bureau. Let's plan on meeting early next week to review all the steps and get the process rolling!

From their list of AM service bureaus, they decided to move forward with Doug and his team. Doug agreed to include team representatives on all build preparation tasks and the build process at the bureau. They will also observe post-processing after the initial build. Roxie and Pete were particularly interested in how quality control was handled in the AM process and looking forward to the upcoming tasks.

REFERENCES

3D Systems (2019). PPI-time zero prints high value reusable packaging with ProJet MJP 2500. www.3dsystems.com/customer-stories/ppi-time-zero-prints-high-value-reusable-packaging-projet-mjp-2500 (accessed 24 August 2020).

Crevier, C. (2020). 3D printing to benefit the dental health of walruses. ÉTS Montréal, www.etsmtl.ca/en/news/2020/impression3d-morses/ (accessed 8 November 2020).

Deloitte Insights (2015). 3D printing for quality assurance in manufacturing. Deloitte Insights. www2.deloitte.com/us/en/insights/focus/3d-opportunity/3d-printing-quality-assurance-in-manufacturing.html (accessed 3 September 2020).

DMG Mori (2020). Lasertec 65 3D hybrid: Hybrid complete machining: Additive manufacturing and milling in one machine. https://us.dmgmori.com/products/machines/additive-manufacturing/powder-nozzle/lasertec-65-3d-hybrid (accessed 24 August 2020).

EnvisionTec (2020). GN ReSound – Hearing aids in 3D. https://envisiontec.com/case-studies/hearing-aid/gn-resound-hearing-aids-in-3d/ (accessed 24 August 2020).

Fabrisonic (2020). Applications: Printing the impossible. https://fabrisonic.com/applications/ (accessed 24 August 2020).

Hanaphy, P. (2020). Canadian researchers 3D print metal tusk caps to protect the dental health of endangered walruses. 3D Printing Industry. https://3dprintingindustry.com/news/canadian-researchers-3d-print-metal-tusk-caps-to-protect-the-dental-health-of-endangered-walruses-178275/ (accessed 29 October 2020).

International Standards Organization (2015). Additive manufacturing – General principles – Terminology. ISO/ASTM 52900:2015(en). https://www.iso.org/obp/ui/#iso:std:iso-astm:52900:ed-1:v1:en (accessed 21 August 2020).

Lockheed Martin (2018). Giant fuel tank sets new record for 3-D printed space parts. https://news.lockheedmartin.com/2018-07-11-Giant-Satellite-Fuel-Tank-Sets-New-Record-for-3-D-Printed-Space-Parts#assets_all (accessed 24 August 2020).

Matsuura (2020). Lumex Avance-60 Metal Laser Sintering Hybrid Manufacturing. https://www.matsuurausa.com/model/lumex-avance-60/ (accessed 24 August 2020).

Mazak (2020). Integrex i-400 AM. https://www.mazakusa.com/machines/integrex-i-400am/ (accessed 24 August 2020).

Porter, K., Phipps, J., Szepkouski, A. and Abidi, S. (2015). *3D Opportunity Serves It Up.* [online] Deloitte University Press, pp. 1–16. www2.deloitte.com/content/dam/insights/us/articles/3d-printing-in-the-food-industry/DUP_1147-3D-opportunity-food_MASTER1.pdf (accessed 9 November 2020).

Rice University (2019). Organ bioprinting gets a breath of fresh air: Bioengineers clear major hurdle on path to 3D printing replacement organs. *ScienceDaily* (May 2). www.sciencedaily.com/releases/2019/05/190502143518.htm (accessed 8 November 2020).

Sciaky, Inc. (2020). Electronic Beam Additive Manufacturing (EBAM®). www.sciaky.com/additive-manufacturing/industrial-metal-3d-printers (accessed 21 August 2020).

Stratasys (2020)b. Connex3 Objet500 qnd Objet350. www.stratasys.com/3d-printers/objet-350-500-connex3 (accessed 21 August 2020).

Ultimaker (2020). Zeiss: 3D printing precision parts for serial production. https://ultimaker.com/learn/zeiss-precision-parts-for-serial-production (accessed 24 August 2020).

Voxeljet (2015). VX4000 The large-format 3D printing system. Voxeljet Data Sheet. https://www.voxeljet.com/fileadmin/user_upload/PDFs/voxeljet_3d-printer_VX4000_2015_EN.pdf (accessed 21 August 2020).

Voxeljet (2017). Creating the SAircraft doors of the future using 3D-printed casting patterns. https://www.voxeljet.com/branchen/case-study/flugzeugtueren-der-zukunft/ (accessed 24 August 2020).

Voxeljet Services (2020). Material data sheet for plastic parts. https://www.voxeljet.com/fileadmin/user_upload/PDFs/material-data-sheet_PPB-PPC_090318_web.pdf (accessed 24 August 2020).

Chapter 6

What Machine and Build Preparation Occurs in Additive Manufacturing?

Case Introduction: Preparing for the AM Build

THE GWM AM Pilot team arrived at Doug's AM service bureau with their STL file on a portable drive, ready for the next steps in the prebuild process. The following is a segment of their conversation:

ED: Okay, Doug! We brought the STL file. Let's fire that powder bed fusion machine up and get this sucker printed! I'm looking forward to a celebratory lunch when this completes printing.

Fundamentals of Additive Manfacturing for the Practitioner, First Edition. Sheku Kamara and Kathy S. Faggiani © 2021 John Wiley & Sons, Inc. Published 2021 by John Wiley & Sons, Inc.

INTRODUCTION

Thorough preparation for the build process is essential to a successful build and the production of a quality part. As discussed in preceding chapters, the selection of material, process, and technology all play a role in producing a part with the desired specifications with the required mechanical properties. These objectives integrate into a complete plan during build preparation to ensure the success of the build. This chapter describes the build preparation steps common to almost all AM materials and processes. It also discusses variations in preparation activities for different AM technologies. Chapter 7 describes the remaining build preparation tasks for finalizing the build plan, addressing quality factors, and performing in situ quality assurance during printing.

AM Notable 6.1

Bringing the Magic to Build Prep

In the early days of AM, build preparation was completed by a team of individuals, including the part designer, manufacturing engineer, and a manufacturing technician or machine operator. They reviewed design parameters, materials, machine specifications, and standard tolerances to make key build decisions such as platform size and position, part orientation, build capacity, part nesting, support structure placement and composition, etc. Lessons learned through trial and error were incorporated into decision-making through experience.

To implement autonomous process flows requiring little human intervention, build preparation software is developing that supports decision-making. Materialise Magics provides data preparation and STL file editor software that converts CAD files to STL, repairs problems, edits designs, and prepares the build platform. Renishaw's QuantAM software tool integrates with its machine control software. It can review build files and help guide the DfAM process accurately. AMFG released build preparation software tools that integrate with their Manufacturing Execution System (MES) and allow designers to visualize and orient parts on the build platform and set optimal build parameters for machines.

Build preparation is another connection for AM as a key element of Industry 4.0.

GENERAL MACHINE AND BUILD PREPARATION TASKS

A fundamental feature of AM is producing a 3D object from a digital representation of the object. In the form of a CAD file, the digital representation is the starting point for build preparation. The steps common to all AM processes appear in bold in Figure 6.1 and are described in this section.

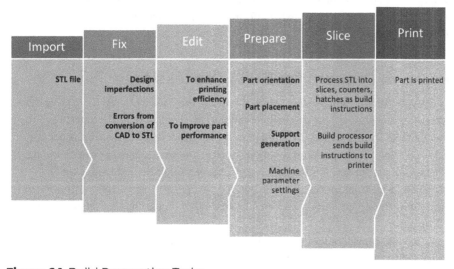

Figure 6.1 Build Preparation Tasks

Source: Original, Kamara and Faggiani, 2020.

Import and Fix the STL File

The final CAD design file of a part or component is converted to STL as described in Chapter 3, and then uploaded or imported to the selected AM system. Generally, the designer converts the native CAD file and forwards it to the machine operator for processing. Further processing of the STL file is required to prepare it and apply technology-specific build instructions for a successful build. STL files are evaluated for errors and modified using standalone STL viewer and editor software or customized, technology-specific software offered as part of an AM system.

Standalone STL viewers provide a platform to verify and validate the file before processing. Since STL files are nonparametric, they are easier to manipulate for Boolean and other operations. Large digital parts and assemblies can be cut and sectioned to allow fitment on the build platform in single or multiple builds.

Customized software used for additional file processing varies by AM system and poses a challenge for the manufacturing technician or operator. Training is essential for the use of build preparation software provided with most mainframe systems. Training, generally structured in phases, focuses on operating and maintaining the specific system. Software training occurs at the OEM's or a subsidiary location, with the hardware training held at the customer's site during and after installation.

The goal of using the STL file build preparation software in this step is to correct any errors in the STL file that may occur as a result of converting the native CAD file to the STL format or issues overlooked in the design process. There are times when the STL file's higher resolution presents problems for the build preparation software, requiring a reduction in the number of triangles. A triangle reduction process is required to reduce the number of triangles and is a feature provided by the software. Some of the customized machine-specific software includes capabilities for sorting and validating the STL file. This step ensures that the coded information for each triangle is adjacent as the Z-height increases.

Edit and Prepare for Build Processing

The AM technology and system selected impacts the choice of materials, layer characteristics, need for supports, and in combination, the mechanical properties of the part printed. In this step of the build preparation process, additional edits occur to the STL file focusing on improving print time and material usage or enhancing mechanical properties. For example, modifications might include reducing or increasing wall thicknesses based on desired part performance. Some software may automatically produce recommendations based on parameters requested by the operator.

Additional considerations that can be addressed by customized software for each AM system and part design include the following.

Part Orientation

Part orientation in a 3D printer can impact the part's strength and accuracy, the number of layers needed, production time, need for supports, and surface finish quality in some AM systems (Redwood, 2020). For example, a cylinder printed with a vertical center axis using FDM would have smoother sides than if printed with a horizontal center axis.

The number of layers required and build time is affected by part orientation. The total number of layers needed to build a tall cylinder is reduced in the horizontal orientation, resulting in a reduction in build time. For part strength of functional parts, technologies like FDM produces parts with anisotropic properties, making them stronger in the XY directions than in the Z direction. An FDM printed part may have four to five times the tensile strength when printed in the XY directions compared to Z. Part orientation is most important in material extrusion.

Part Placement

The placement of the part on a build platform in an AM system impacts the accuracy of the part build, build time, and total costs. Each machine has an available build volume, and the challenge involves how to best "fit" the number of parts to be printed into the build chamber to maximize the output per build, thus saving time and cost (Araujo et al., 2015). Again, customized software can produce suggested build patterns to accommodate specified constraints and objectives.

In powder bed fusion and vat photopolymerization systems, where the laser position is at the center of the build platform, parts placed at the center of the build volume are generally more accurate with improved properties than those positioned away from the center. Recent innovations in using multilaser systems in powder bed fusion systems, like EOS GmbH's EOS M 400-4 and SLM Solutions' SLM®-500, each with four lasers located in each quadrant of the build platform, minimizes this effect while increasing throughput (EOS, 2020).

Support Structures

The need for support structures in AM depends on the process, material, and part design. Supports may be needed to prevent deformation or collapse of features during the printing process, especially when an overhang or angle is part of the design. Parts not yet connected to the main body of a printed part may require supports as the fabrication proceeds, or support to balance a part and securely connect it to the build platform may be needed. Also, support structures can help reduce thermal distortions that may lead to shrinkage, cracking, curling, delamination, or sagging during some processes.

Binder jetting and polymer-based powder bed fusion allows the operator to upload multiple files within the build platform and provide the ability to apply individual machine build parameters since support structures are not required. Technologies that use supports require the analysis of individual part files, as support structures may be critical for successful builds. Individual STL files can be combined before support generation by the machine software to reduce processing times. Support structures add extra material to a build, which means more time and cost. In material extrusion and jetting, the support material is different from the model material and generally water dissolvable. In metal powder bed fusion and vat photopolymerization, part material also creates the support structure. Specialized software can suggest support structures that minimize material usage while providing adequate strength to produce a more optimal build, ensuring part accuracy and integrity.

Slicing is the process of generating information for each layer of the part. For operations requiring supports, like material extrusion and jetting, metal powder bed fusion, and vat photopolymerization, the digital part is presliced before build instructions are transmitted to the machine. These systems do not allow an operator to add additional parts after the build has started since every part needs to be attached to the platform. In systems that do not require supports, parts are sliced on the fly during the build process. This approach allows additional parts to be added during production, assuming there is enough material to complete the entire build. Parts are free-floating within the build volume and supported by unfused or unsintered or unmelted powder.

Build Platform

The build platform is a critical part of the prebuild process that needs careful attention for technologies like metal-based powder bed fusion, material extrusion and jetting, and vat photopolymerization since the leveling, flatness, and cleanliness of the platform can affect part quality. Solvents can completely remove supports by softening or dissolving solidified material on the platform from a previous build in some vat photopolymerization machines. However, completely removing any solvent residue is important to ensure the existing resin vat is not contaminated. In metal-based powder bed fusion, support structures and sometimes parts are welded directly onto the platform. With a layer thickness of around 20 microns, the build platform's flatness is critical for a successful build. Generally, processes that do not require the use of supports do not use removable build platforms.

Machine Calibration

All AM technologies require precalibration checks. For example, even a low-cost material extrusion system extruding plastic filament needs to verify the extrusion tip's desired temperature before building parts. A mainframe vat photopolymerization system verifies the laser power to ensure power is adequate for the build. Resin levels need a review to ensure a sufficient amount of photopolymer to start the process. Depending on the system and technology, some calibration tests can be manual or automated. Equipment manufacturers typically provide training and requirements for prebuild machine calibrations.

Part orientation, part placement, support structures, build platform, and machine calibrations get addressed during the build preparation phase. The following section discusses technology-specific build preparation considerations.

AM Notable 6.2

Calibrate for Quality

Calibration is one of many aspects of making high-quality parts consistently. The qualification of parts and processes for acceptance in major industries served by metal AM rely, in part, on calibration. In short, calibration indicates a good manufacturing process by ensuring a measured parameter correlates to some agreed-upon value. AM is still in the early stages of maturity concerning industry-specific standards, though organizations like NASA have established 3D-printed part quality requirements that are flowing into newer standards.

It was not feasible on early metal AM machines to perform a calibration before each build, forcing a choice between quality assurance and production efficiency. Now a prebuild calibration is a necessity and is supported by the newest metal AM system capabilities. For example, calibrating laser optics once required the system to be taken offline, making it difficult to recalibrate in the middle of production. Next-generation systems offer prebuild features that streamline and automate calibration so machines can stay online for calibrations (Murphree, 2020).

In metal powder bed fusion systems, calibrating the platform is key to part quality, where exactness of both thickness and uniformity of the powder bed is needed for layer to melt properly under laser contact. NASA requires powder bed qualification every 180 days, though six months leaves ample time for the recoating process to go wrong even though bed images are periodically reviewed. Next-generation powder bed fusion systems can check the recoater blade before, during, and after every build using height-mapper metrology systems that measure powder bed topology. Specific quantitative measurements ensure the powder layer delivered by the recoater is within specifications for uniformity and thickness.

Calibration of lasers, powder bed quality, and other factors is essential before a build to identify and prevent problems that result in build failure or poor part quality. As more AM machines adopt sophisticated, automatic calibration and measurement features, part quality should become even more consistent and predictable.

AM PROCESS-SPECIFIC MACHINE AND BUILD PREPARATION

Machine and build preparation tasks will vary by AM process and material. A summary of different preparation tasks and considerations appears in Table 6.1. The following section describes build preparation activities related to materials, platform preparation, support structures, and energy and throughput settings.

Table 6.1 Pre-Build Considerations by AM Process

AM Technology	Material	Platform	Support Structures	Energy/Throughput
Vat photopo-lymerization	Resin age and viscosity impact part quality	Build orientation critical to achieve desired part properties.	Supports needed for angles < 45 degrees and overhangs exceeding 0.4".	Determine the critical exposure, hatch spacing and overcure parameters for optimal build.
Material jetting	Different materials are used for part vs support. Material age and expiration date need to be monitored.	Removable, washable build platform. Build orientation key for good resolution and support minimization.	Support material volume may exceed part volume, and part orientation, support, and material volume are interconnected.	Nozzles may require preheating at time of printing.
Binder jetting	Powder shape and size, binding agent droplet size impact part properties.	Build orientation used to maximize throughput and reduce build time	Not required	Binder saturation and drying time affect Z-growth and part quality.
Material extrusion	Materials must be stored in sealed container during/after build to prevent moisture absorption.	Build orientation used to maximize throughput.	Always required	Dual or multiple extrusion heads used for model and support materials.

Table 6.1 *(Continued)*

Powder bed fusion	Powder size, distribution, and composition impact part properties. For metal, powder layer thickness should exceed particle size.	Preheated build platform required to prevent warpage and assure accuracy. Proper build orientation for polymers can improve part strength or flexibility.	Polymer-based materials do not require supports and can be built on top of or inside each other. Metal materials require supports and parts cannot be stacked on Z-axis.	Laser power and scanning space used to manipulate energy density, with higher energy density resulting in stronger parts.
Direct energy deposition	Powder size and wire diameter impact part properties.	Build can occur on an existing part or platform.	Cannot produce supports, but use of multi-axis extrusion heads enables complex geometries.	Laser, electron beam, plasma or electric arc produce high density parts. Fast powder distribution and wire filament delivery rate result in high throughput.
Sheet lamination	Long-fiber fabrics, nylon, thermoplastics integrated with carbon or glass fibers used.	No specific requirements for build platform.	None required	Not applicable

Source: Original Kamara and Faggiani, 2020.

Vat Photopolymerization

In vat photopolymerization, the photopolymer's viscosity is a factor in a successful build process. Photopolymer's viscosity increases or thickens over time due to laser scatter and exposure as it moves across the platform, curing the part profile. New replacement resin is typically added before each build, substituting the previous build's used resin volume. Because of the aging effect, the ratio of new resin to old resin in the supply tank for a build is significant. For example, a large frame SLA machine may hold up to 65 gallons (15,000 cubic inches) of resin. Part volume from a build may use only 50 cubic inches, less than half a percent. Certain resins may require replacement of the vat of resin annually based on usage and change in viscosity from new material. The age of the resin and viscosity should be checked regularly and adjusted as needed.

Part orientation in vat photopolymerization processes is critical as it affects part quality, build integrity, and throughput. Support structures are necessary for this process. Operators must ensure that hanging surfaces are supported. Angled surfaces of less than 45 degrees and overhangs less than 0.4 inches do not require support structures. Operators can increase the throughput of multiple parts if all parts have approximately equal Z-height. There is additional Z-layer growth when numerous layers cure at once due to the laser's penetration depth. For example, a cylinder orientated sideways may be slightly elliptical compared to a vertical orientation that ensures a circular and accurate profile.

Excitation characteristics of the resin determine the correct laser scan speed in laser-based vat photopolymerization systems. These characteristics represent the energy required to change the resin from liquid to solid. The amount of energy delivered is a function of the laser power and scan spacing of the laser path as it traverses the cross-section of the part. Scan spacing is the distance between the centers of adjacent laser paths. For DLP-based vat photopolymerization systems, the entire layer is flashed or fused at once with a higher resolution, typically, 1024 × 780. Some DLP-based VP systems build parts bottom-up. These systems eliminate laser penetration depth issues since light is flashed with a controlled layer onto a transparent platform holding parts up from the bottom, not into a vat filled with resin. Figure 6.2 illustrates parts built with vat photopolymerization.

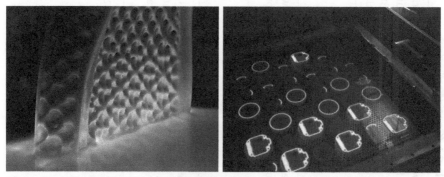

Figure 6.2 Vat Photopolymerization Builds
Source: Stratasys Ltd., https://www.stratasysdirect.com/manufacturing-services/3d-printing/3d-printing-costs-ways-to-save-increase-efficiency

Material Jetting

Material jetting utilizes different materials and separate nozzles for the part and support materials. For photopolymer-based materials that are mostly acrylate-based, the heat deflection temperature is generally lower than that of epoxy-based resins in vat photopolymerization. It is essential to ensure these materials are used by their expiration date since acrylate-based photopolymers are susceptible to thickening and have a shorter shelf-life. Most systems that require a barcode scan to load materials will reject material with an expired date that is 12–18 months past, depending on the material. Wax-based materials have a longer shelf-life and are used primarily for their low melting temperature. Investment casting or lost-wax processes for jewelry applications make use of these materials.

Build orientation is a key factor in the material jetting process for material resolution since support structures are required coupled with a higher Z-resolution. For example, Stratasys' Objet350 and Objet500 Connex3 have an XY-resolution of 600 dpi compared to a 1600 dpi Z-resolution. Based on part geometry, support structure material volume may surpass part material volume in the build. Part orientation will affect support design, which eventually affects lead-time and overall cost of the part. The support materials can be water dissolvable or easily removed with a water jet. Figure 6.3 shows build orientation for 3D printing with wax.

Material jetting systems utilize a removable platform that is easily washable and reusable since the supports are generally water dissolvable. Like all photopolymer-based materials, they are susceptible to humidity; however, the material stores in sealed containers are not exposed to environmental conditions.

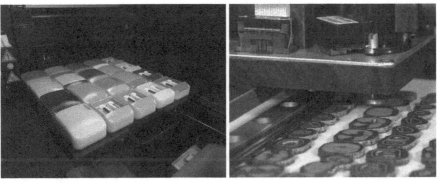

Figure 6.3 Build Orientation – Wax 3D Printing
Source: Moreno Soppelsa/Shutterstock.

Binder Jetting

Powder shape, size, and distribution play a key role in binder absorption in binder jetting systems. Drop-on-demand printheads deposit picoliter-sized droplets of a liquid binding agent onto a powder bed surface in binder jetting. The velocity of the droplets as they hit the surface impacts spreading and absorption dynamics, which can impact the part's quality and integrity, which can also be affected by powder characteristics (Miyanaji, Momenzadeh, and Yang, 2018). Figure 6.4 shows various powders used in binder jetting.

Since no supports are required in binder jetting, build orientation is used to leverage throughput and build time. Parts may be nested together within the build volume without sacrificing part quality. For sand-casting applications, parts generally ship with a custom support structure surrounding them to minimize damage.

Binder jetting binding agents are important to the process as they offer unique benefits to printed materials. Typically, droplets in sizes between 40–60 microns impact powder with velocities ranging from 1–10 m/s. These binders can be inorganic or water-based, with furan or phenolic resins used for foundry applications. Size of the droplets, drying time, droplet velocity and spacing, impact dimensional accuracy, porosity, and strength of the final part (Colton, Liechty, McLean and Crane, 2020).

Figure 6.4 Binder Jetting Metal Powders
Source: MarinaGrigorivna/Shutterstock.

Material Extrusion

Most materials used in material extrusion processes are hygroscopic with an affinity to absorb moisture from the environment. Mainframe systems store material in a sealed container during and after a build to minimize absorption. Moisture in filament-based materials generally affects the material's melting or softening temperatures, resulting in lower mechanical properties. The polymer is heated at its softening temperature enabling fusing at the previous layer or platform. Parts made from low-temperature materials, such as PLA, are built on an open platform at room temperature. Most of these systems include a heated platform to prevent warpage and promote adhesion for the first few layers of the part and supports. For high-temperature polymers, like polycarbonate (PC) or polyphenylsulfone (PPSF), a heated and enclosed build chamber is required for controlled airflow to produce consistent mechanical properties.

Figure 6.5 Support Structures Used in Material Extrusion
Source: Stratasys Ltd.

Several systems are now available that utilize material in pellet form, used in injection molding. Pellets offer low-cost access to a wide range of materials. Materials in pellet form are cheaper compared to their counterparts in filament form. However, given their hygroscopic property, care must be taken to store pelleted materials away from moisture until added to the machine before the build.

Material extrusion always requires support structures due to gravity since the material is deposited vertically onto the platform. Proper part orientation can maximize throughput and need careful planning for multiple parts printed simultaneously. Large mainframe systems typically have dual extrusion heads dedicated to part and support materials, which are different. Supports materials can be water-soluble or breakaway, depending on the model material. Most low-cost material extrusion systems use a single extrusion head, and the same material used for both parts and support structures. Figure 6.5 displays a printed part with support structures.

Powder Bed Fusion

Powder size generally varies between a 50- and 90-micron particle for successful parts. The material is generally not 100% recyclable, though some excess may be recoverable. A composite material that incorporates glass and other fibers is especially critical since the ratio of polymer to glass or fibers could not be easily maintained during recycling resulting in inferior mechanical properties. Large mainframe systems with automated material handling systems adjust the ratio of new and used material to produce consistent parts.

Sacrificial parts or tensile bars are placed at the bottom of each powder bed fusion build to prevent warpage and increase the surface temperature before printing starts. Melting or sintering parts before the warmup process results in part warpage and inaccuracy. Directional differences are minimized in polymer-based powder fusion since multiple layers are cured repeatedly in the Z-direction, exposing the layers to higher energy. Parts can be close in achieving similar mechanical properties in all directions, XYZ. This additional energy enables the strength in Z-direction to be slightly higher but closer to *XY* to create a part with uniform properties in all directions

Polymer-based parts built with powder bed fusion need no supports and may be positioned on top of each other during the build without sacrificing mechanical properties and quality. A build chamber temperature set to 10–16 °F below the polymer-based powder's melting point is ideal. Part strength is typically higher along the Z-axis due to overcure. If a component design includes a living hinge for additional flexibility, the living hinge must be oriented and built along the *XY*-axis for the best performance. Figure 6.6 displays how parts can be nested to take advantage of full capacity of a build chamber.

The total amount of energy imparted on the powder is directly related to the strength of the parts. The energy level is achieved by increasing the laser power or reducing the scan spacing (spacing between each laser pass) to increase the surface's energy density. An understanding of this relationship allows the practitioner to leverage the capabilities of AM. For example, producing parts with increased scan spacing will produce a porous part with a lower density using the same machine and material. Part mechanical properties and performance requirements should drive these machine settings.

Optimum powder size for metals generally varies between 15- and 45-microns for laser-based systems and higher for electron beam-based systems. Studies show that part density is affected by the powder's particle size distribution, impacting its ability to flow and spread. For the best result, optimal powder layer thickness should be larger than particle size, corresponding to the powder's 90th percentile (Colton, Liechty, McLean,

Figure 6.6 Multiple Nested Parts Built Using Powder Bed Fusion
Source: Courtesy of 3D Systems, Inc.

Figure 6.7 Support Structures Required in Metal Powder Bed Fusion
Source: MikeDotta/Shutterstock

and Crane, 2020). Gas and plasma atomization produce the powders since both achieve a more spherical shape preferred in AM systems.

Powder bed fusion systems require support structures for metals. Metal parts cannot be stacked in the Z-axis like polymer-based parts due to the support structures interfering with the other parts. Due to internal stresses, while melting the layers, support structures must hold the melted parts onto the platform, preventing the re-coater blade from hitting parts. Figure 6.7 shows a part with support structures printed using powder bed fusion.

In powder bed fusion systems using metal, most of the energy for melting powder comes from the laser. Fiber lasers with 200W+ are standard. For electron beam systems, like GE's Arcam EBM Q20plus, power is variable up to 3000 W with scan speeds of up to 8000 m/s compared to 7 m/s for laser-based systems ('EBM_QPlus20_Bro_A4_EN_v1.pdf', 2020). With this capability, electron beam systems preheat the entire part bed before melting the parts. Electron beam-based systems generally have a higher density compared to laser-based systems.

Directed Energy Deposition

Directed energy deposition utilizes powder or wire feedstock based on the system. Powder size generally varies between 50- and 150-micron and wire diameter between 1–3 mm. These systems have the highest throughput for direct metal systems in additive manufacturing. These systems will produce any material in powder or metal form that is weldable (*Digital Alloys*, 2019). Systems with multi-nozzles are capable of welding multiple materials or increased feed rates with the same material.

Directed energy deposition is ideal for repairing existing parts since metal deposition can occur on an existing part or platform. The DED process is incapable of producing support structures, limiting specific geometries, although its multi-axis extrusion heads make up for this limitation. Systems like Sciaky's EBAM 300 machine are capable of producing parts up to 19 ft long (*Metal Additive Manufacturing Systems*, 2020)

DED's energy source derives from an electron beam, laser, plasma, or electric arc, and the parts are highly dense. Most systems that are not hybrid produce near net shapes that are post machined to achieve design tolerances.

Sheet Lamination

As of October 2020, Composite-based Additive Manufacturing (CBAM) technology, developed by Impossible Objects, is the only sheet lamination technology on the market that is not a hybrid (Impossible Objects, 2020). The material used is in the form of long-fiber sheets of carbon or fiberglass that feed into a 3D printer and printed on with a thermoplastic matrix material ("Materials," 2020). The powder is applied to the sheet and adheres to the liquid material.

After removing excess powder, the sheets are heated to the thermoplastic material's melting point and then compressed to achieve the desired part dimensions. Blasting removes the unbonded sheets. Prebuild processes for CBAM involve replenishing materials, determining the placement of sheets on the build platform, and heating the roller used to add pressure to the sheet layers.

Ultrasonic additive manufacturing (UAM), the other primary form of AM sheet lamination, uses ultrasonic vibrations under pressure to fuse metal foils. A base plate fixes into the machine, and metal foil feeds under a sonotrode that produces the ultrasonic pulses and applies them to the foil. Another AM process, usually CNC machining, is used to generate the desired shape from the ultrasonically consolidated material. Prebuild procedures include material loading and verifying the calibration of sonotrode devices.

This chapter described factors involved in preparing for the build process and factors related to achieving a quality result from the build process. Upfront planning and consideration of build orientation, build platform preparation, part placement, support requirements, material requirements, quality, and energy source and throughput, will help ensure a quality part results from AM. Chapter 7 describes additional prebuild tasks.

AM Notable 6.3

Build Preparation for Bioprinting

Pre-bioprinting steps differ significantly from the build preparation steps required for other AM processes, though they share some common purposes. AM processes require a design file that can be translated to the STL format readable by most AM technologies. In bioprinting preparation, a model is likely created from a biopsy of an organ using computed tomography (CT) or magnetic resonance imaging (MRI). 2D layer images sent to the 3D printer are created by a tomographic reconstruction process, similar to the slices generated from an STL file.

Material preparation for bioprinting is more involved than for other AM processes. After the image to be bio-printed is created, selected human cells are isolated and multiplied, then combined with a special liquefied material that provides nutrients and oxygen to keep them alive. If the cells are part of an extrusion-like process, they are first encapsulated in cellular spheroids. The liquid blend of cells, matrix material, and nutrients is known as *bioink*. The bioink is placed in a cartridge in preparation for deposition based on the 2D images (Javaid and Haleem, 2019).

Bioinks are suitable for a variety of different 3D printing processes, including photolithography, magnetic 3D bioprinting, stereolithography, and direct cell extrusion. Biomedical AM applications are a significant growth area in AM, and ethical issues related to their use are slowly being addressed.

Case Conclusion: Build Preparation Summary

The GWM AM Pilot team worked with Doug to prepare their original STL design file for the AM system's 3D printer. After completion of most prebuild tasks, the following conversation ensued:

ED: That was a lot more involved than I thought it would be.

PETE: Look at it this way, Ed – if it replaces all the retooling we'd need to do in our current operation to implement the newly designed part, it's definitely worth it!

BOB: Of course, *if* our traditional methods could even manufacture the new design!

FRANK: True, Bob. Though less retooling sounds good to my ears! That could only mean lower costs for implementing part redesigns or new parts and products.

ROXIE: From a quality assurance standpoint, making sure the AM system machines are correctly calibrated and build orientations are optimized to ensure part properties makes a lot of sense. My concern is the introduction of variation in AM processes that will negatively impact quality! Where do we go from here, Doug, on those significant quality factors?

DOUG: Glad you asked, Roxie! We're ready to move on to the build process!

REFERENCES

Colton, T., Liechty, J., McLean, A. and Crane, N. (2020). Influence of drop velocity and droplet spacing on the equilibrium saturation level in binder jetting. In *Solid Freeform Fabrication 2019*. Proceedings of the 30th Annual International Solid Freeform Fabrication Symposium. [online] pp. 99–107. http://utw10945.utweb .utexas.edu/sites/default/files/2019/008%20Influence%20of%20Drop%20Velocity%20 and%20 Droplet%20Spacing%20on.pdf (accessed 9 November 2020).

EBM_QPlus20_Bro_A4_EN_v1.pdf (2020). www.ge.com/additive/sites/default/ files/2020-07/EBM_QPlus20_Bro_A4_EN_v1.pdf (accessed 29 September 2020).

EOS. (2020). M 300 Series. Digital additive manufacturing with metals. www.eos. info/en/additive-manufacturing/3d-printing-metal/eos-metal-systems/eos-m-300-4 (accessed 29 September 2020).

Impossible Objects (2020) The Printer. *Impossible Objects*. www.impossible-objects.com/ cbam-printer/ (accessed 8 October 2020).

Javaid, M. and Haleem, A. (2019). 3D printed tissue and organ using Additive manufacturing: An overview. *Clinical Epidemiology and Global Health*, 8. doi: 10.1016/j. cegh.2019.12.008.

Materials (no date). *Impossible Objects*. www.impossible-objects.com/materials/ (accessed 29 September 2020).

Metal Additive Manufacturing Systems (2020). www.sciaky.com/additive-manufacturing/industrial-metal-3d-printers (accessed 29 September 2020).

Miyanaji, H., Momenzadeh, N. and Yang, L. (2018). Effect of printing speed on quality of printed parts in binder jetting process. *Additive Manufacturing*, 20, pp. 1–10.

Murphree, Z. (2020). Why pre-build calibration is critical to part quality for metal additive manufacturing. *Quality Magazine* (Oct 5). www.qualitymag.com/ articles/96242-why-pre-build-calibration-is-critical-to-part-quality-for-metal-additive-manufacturing?v=preview (accessed 8 November 2020).

Redwood, B. (2020). How does part orientation affect a 3D print? *3D Hubs*. www.3dhubs. com/knowledge-base/how-does-part-orientation-affect-3d-print/ (accessed 28 September 2020).

Chapter 7

What Occurs During the Additive Manufacturing Build Process?

Case Introduction: Estimating Build Time and Ensuring Quality

Before the actual build began, Doug took some time to address AM quality and explain some steps that the service bureau would be taking on the GWM AM Pilot team's part. Part of the conversation follows below:

DOUG: AM systems and technology developed so rapidly that it shot out ahead of quality assurance and parts qualification. Much is happening in AM on these fronts, and new methods for quality assurance are appearing.

ROXIE: Our discussion about verifying the design and material properties made a lot of sense in terms of part quality, but I'm still concerned about process variability and its impact on part quality.

Fundamentals of Additive Manfacturing for the Practitioner, First Edition. Sheku Kamara and Kathy S. Faggiani © 2021 John Wiley & Sons, Inc. Published 2021 by John Wiley & Sons, Inc.

PETE: I read something about models and simulations of the build process, Doug. Might those approaches address some of Roxie's concerns?

ED: Well, won't our existing destructive testing approach work? Why do we have to change our backend processes?

FRANK: Our existing approaches are relatively expensive and time-consuming when you think about how many parts we build, destroy, or throw away.

BOB: Plus, the testing doesn't always tell us what we need to know and may not be accurate, to add insult to cost injuries!

DOUG: Let me walk you through quality in the AM world, and maybe I can address your issues. Plus, Ed, I think we'll close in on the celebration time frame by generating an estimate of when your first part will roll off the printer. That should help facilitate the party planning!

INTRODUCTION

The build process, where all the action happens, is likely to require the least amount of attention as a part of the AM approach. Occasionally some systems, such as material extrusion or material jetting, may require an operator to replenish materials during the build process. For example, the Stratasys F900, a mainframe machine with a 36 × 24 × 36 inches platform, holds a maximum of 184 cubic inches of model and support material separately. If the combined volume of the part(s) to be printed exceed this maximum, the operator needs to change canisters during the build without interrupting the build process. Given AM processes and systems' maturity and stability, the build process typically runs without operator monitoring or intervention. Freeing up operator time and attention to focus on other tasks helps enhance the overall manufacturing productivity gains.

Occupational health and safety, along with quality control and assurance, are two areas of concern within the AM build process that warrant further attention. Quality remains a struggle for many additive manufacturing practitioners, and innovative approaches to monitoring the build process and providing real-time feedback are under development. Another area of concern for AM practitioners during the build process is occupational health and safety, given the range of new materials, methods, and technologies introduced in AM. This chapter focuses on the build process and quality in AM, starting with an overview of the build process and estimating build time, then moving on to quality in AM and addressing quality before and during the build process. Included is a discussion of potential safety concerns and software available to facilitate prebuild calculations and monitor quality.

QUALITY IN AM

Quality assurance has been viewed by many as the most significant obstacle to widespread adoption and implementation of AM, particularly for metals. The primary components of AM quality have been categorized and are depicted by Deloitte in Figure 7.1 and discussed briefly below (Wing, I., Gorham, R. and Sniderman, B, 2015).

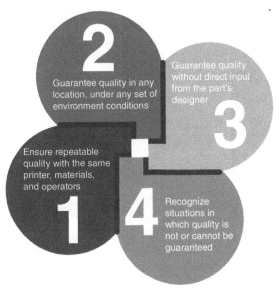

Figure 7.1 Key Components of AM Quality
Source: Wing, I. et al. 2015 / Deloitte University Press.

Repeatable and Reproducible Quality

Repeatability relates to the precision of "independent test results [that] are obtained with the same method on identical test items in the same laboratory by the same operator using the same equipment within short intervals of time" (Dowling et al., 2020, p. 2). Reproducibility is defined similarly to repeatability. Rather than achieving the same results under identical conditions, tests may occur in different laboratories or locations, with different operators, other environmental conditions, and various equipment to demonstrate reproducibility. The aspects of repeatability and reproducibility are critical to demonstrate quality in AM. AM quality must be high for it to continue to grow across industries.

A variety of factors in AM can impact repeatability and reproducibility. For example, in powder bed fusion, the material and machine characteristics, such as variation in powder morphology or equipment calibration, are the primary focus in ensuring repeatability during prebuild activities. During the build itself, laser interaction with the powder bed becomes the most important factor, particularly energy density or specific energy input interactions with powder density, beam diameter, and interaction time. Any variation in these characteristics, such as beam focus moving out of alignment, can directly impact the build quality and represent a critical factor in repeatability (Dowling et al., 2020).

In another example, since temperature and humidity fluctuations impact the materials and integrity of the parts produced by material jetting and vat photopolymerization, these processes require a controlled environment. Technologies that utilize an electron beam

Figure 7.2 GE Arcam Vacuum Chamber
Source: General Electric Company (GE), https://www.ge.com/additive/ebm.

typically melt powder or filament in a vacuum build envelope isolated from the general environment, as shown in Figure 7.2. Only materials require protection from the general climate in other AM technologies.

Independent Quality

Ensuring that equipment settings related to energy sources, build environment, and other build characteristics are correctly established and maintained throughout the build process are essential to a quality outcome. OEMs continue to struggle with the amount of parameter control they are willing to grant to operators and, at the same time, guarantee that the parts are within specifications. During the early years of AM, operators controlled almost every machine and material parameter to produce acceptable parts. As the industry has matured, machine manufacturers have limited or licensed the material used in the systems, and the amount of machine parameter freedom is now limited.

For example, binder jetting and material jetting offer minimal opportunities to affect part quality since operators can change only a few parameters. Some material extrusion, material jetting, powder bed fusion, and vat photopolymerization equipment utilize an RFID reader to update material and machine parameters automatically with the scan of a material canister at loading time. Except for the environmental conditions, these systems can produce consistent parts irrespective of operator experience.

As discussed in Chapter 3, one note of caution is that part orientation influences the part quality. Hence, decisions about part orientation are a crucial variable under the control of the designer and operator. AM design intent may be interpreted differently by different operators, and the resulting circumstances may be different. Consider two operators producing the same part. In one of the systems, a single part needs printing, while in the other system, multiple parts await printing. Decisions about part placement made by the operators influence part quality.

Fortunately, many of the key machine parameter settings are determined empirically by the software that optimizes build processes for the specific AM system that receives the prepared STL file. The software can also balance competing priorities to optimize build orientation and build platform part placement and provide the most efficient and effective support structure design.

Non-Guaranteeable Quality

The need to identify and acknowledge situations in which AM quality is not, nor cannot be, guaranteed is the essence of non-guaranteeable quality. Variability is inherent with AM systems, as with other types of manufacturing systems. In most AM systems, components located in the center of the build platform are more accurate and consistent than those placed away from the center. In single laser-based systems, for example, powder bed fusion, as the laser scans away from the center of the platform, the beam profile changes from circular to elliptical. Figure 7.3 shows a laser moving away from the build center. Multi-laser systems focusing on certain quadrants of the build platform minimize

Figure 7.3 Laser Scanning Away from Build Center
Source: Moreno Soppelsa/Adobe Stock Photos.

this effect. For systems that require heated build platforms or chambers, parts closer to the platform's edge tend to be at somewhat lower temperature than those at the center of the platform. In a production environment, operators can produce multiple builds to dial in the parameters and achieve consistency.

AM Notable 7.1

"Hands On" Quality at EOS

How do you ensure uniform, repeatable, and reliable AM printed parts?

EOS, a worldwide leader for industrial 3D printing of plastics and metals, has developed a customer-centric approach to AM quality. Their quality assurance process controls the quality of every part at every step in the value chain. They create the conditions for uniform, repeatable, and reliable part properties by coordinating among systems, materials, and processes (EOS, 2020).

EOS systems have a rigorous machine acceptance process in which system testing uses common ISO and DIN standards. They manufacture a predefined reference object on each system and verify it against criteria including surface quality, porosity, tensile strength, and elongation at break. They also conduct a comprehensive check of all system components.

EOS is renowned for its powder-based systems. Multiple testing procedures check the chemical properties and uniform particle size distribution of raw powders. Density cubes and tension rods are manufactured from each powder batch using fixed criteria. Processing and packaging of powders only occur after successful testing. EOS further classifies the technology maturity of their metal and polymer materials and processes using the NASA Technology Readiness Levels (TRLs).

AM experience and extensive testing allowed EOS to establish all parameters for achieving specific chemical properties or part properties, such as layer thickness or laser power. Material suppliers and hardware development teams coordinate to optimize the potential of technology, materials, and process, resulting in the optimal combination of AM parameters to develop a specific product property.

The EOS story illustrates one model for AM quality assurance.

BUILD PLANNING AND THE BUILD PROCESS

The build planning process consists of activities involved in developing a machine plan to produce a specific part. Chapter 6 describes the tasks involved in importing, fixing, editing, and preparing for AM. Figure 7.4 summarizes these tasks, and the steps to be discussed in this chapter appear in bold. This chapter also includes a discussion of quality assurance functions important in this phase.

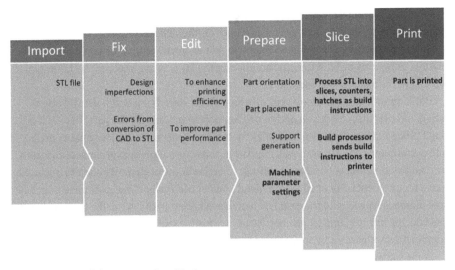

Figure 7.4 Build Preparation Tasks
Source: Original, Kamara and Faggiani, 2020.

Part geometry, surface finish, and mechanical properties identified before the build, along with the selected material, process, and technology, have been determined. The CAD design file, converted to STL (or another format) compatible with the AM system, enables 3D printing. In this step, the goal is to identify and fix design imperfections and errors introduced in the conversion to STL to help avoid build failures and ensure a high-quality part. Editing may also occur to ensure printing process efficiency, optimize material volume, or improve part performance or properties.

Identifying and verifying the build parameter settings on the AM machine to fulfill the required part specifications becomes the new focus. With this information settled, build time can be reliably estimated.

While some AM applications may move on to printing one or more practice parts to verify build parameters, this is not always feasible due to material and AM system costs. Sophisticated software, sometimes included in an AM system package, is increasingly used for this purpose. The following sections discuss the modeling and simulation of the build and estimating build times.

Modeling and Simulation

Although part geometry and build orientation are key factors in building planning, various parameters impact the build's success in achieving design specifications and quality requirements. These parameters might include energy level, scan speed, build chamber temperature, and many others, some of which are AM system-specific. Advanced analytical models exist to simulate the physical phenomena occurring within AM processes and optimize parameter settings to achieve desired outcomes. Table 7.1 lists several popular software modeling tools and summarizes their characteristics.

Simulation software is defined broadly in AM as software that focuses on design validation, improving build setup, and simulating the 3D printing process, the latter two of which are the focus of this section.

Build optimization software typically focuses on identifying optimal build orientation and part placement and optimizing support structures for the part. In contrast, build process simulations help optimize multiple parameters to achieve quality goals and performance objectives. Build simulation software supports experimentation with a broad combination of parameter settings to arrive at the ones most suited to a specific part's requirements. For example, in metal powder bed fusion, laser parameters include spot diameter, speed, and the distance between adjacent paths. Also, the laser parameters used during printing impact geometries of part features and supports, and only certain combinations of parameter settings are likely to produce high-quality parts.

Some software, such as Renshaw's QuantAM, supports using a spreadsheet to generate parameter sets for different test parts and then produces the build instructions for the 3D printer (Zelinski, 2020b). Like Materialise Magics Simulation Module, other packages allow the operator to optimize a virtual build platform to reduce the risk of print failures without printing test parts and identify the best parameters to build a part. The software allows the operator to analyze how the digital imitation will behave during production (Materialise Magics Modules, 2020).

The number of variables for simultaneous consideration and the available computing capacity place limits on simulations. Some researchers suggest that a fully representative AM simulation requires over 130 different variables (Pal et al., 2013). Accurate simulations are essential for high-risk situations to ensure defect avoidance and guarantee quality. In these scenarios, simulation models run at high-performance computing facilities, or cloud-based computing environments ensure adequate computing capacity. Less risky applications may rely on simulations with fewer variables that represent part design specifications and requirements.

Estimating Build Time

The build time for the entire build process may take anywhere from minutes to days. The build process time varies based on AM technology and the size of the build. Many AM systems require a heated build plate or preparation of the build envelope environment before the start of 3D printing. Some AM processes require heat treatment for stress relief during the printing process. The overall cycle time includes actual 3D print time and the time required for these other activities.

The build envelope, or maximum area for part production on an additive manufacturing system, and the envelope's use are key factors in build time estimation and costing of AM parts. A part is built to the build envelope's size unless the part is divided into sections, with sections built separately and joined after build; a part that cannot sub-

Table 7.1 AM Build Simulation Software

Name/Company	Description	Key Features
Materialise Magics Simulation Module	Creates visual prototypes and allows control of part quality and reduction of part failure for metal AM	• Used with Magics AM system • Fast/robust simulation on desktop computer • Easy to use, no simulation expertise required • Can separate visualization of support from part

Simufact Additive MSC Software	Optimizes powder bed processes (SLS, SLM, LBM, DMLS, EBM) examining influence of: material selection, power vs. speed, powder characteristics, build path/hatching patterns, support/internal structures	• Computation methods can predict:
		  • Shrinkage
		  • Warpage
		  • Residual stresses
		  • Optimal location of part
		  • Support structure
		  • Deformation
		• Runs in Windows or Linux environments

AutoDesk NETFABB	Predicts thermomechanical response of additive parts during PBF and DED	• Physics-based, multiscale for low processing time and high accuracy • Predicts distortion/support failure • Simulate entire build plate
ANSYS Additive Science	Simulates metal AM processes; determines best process parameters for any machine/material combination	• Distortion compensation/stresses • Build failure prediction • Full user access to process settings
Additive Works Amphyon	Supports LBM processes (LBM, SLM, DMLS, metal 3D printing)	• Optimize build orientation • Optimize support structure • Simulate mechanical AM process and calculate residual stress and distortion fields • Thermal process simulation

Source: Original, Kamara and Faggiani, 2020.

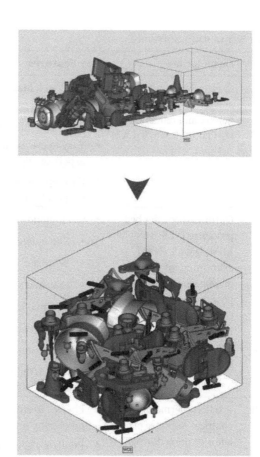

Figure 7.5 Parts Placed within Build Platform by Software
Source: Image of 823 parts nested on an EOS P700 platform with the Sinter Module of Materialise Magics 24.

divide with sections joined after printing requires a larger build envelope. The build envelope can maximize build capacity by printing multiple parts in the same build process. By maximizing build envelope capacity, it is possible to use less energy during the build process, save on build time, and reduce per-part costs.

Build time estimation is complicated, as it is specific to the additive manufacturing system and material used. Part orientation and material volume also influence build times, as do a variety of other factors. Estimating build time is an important factor in planning for production and estimating production costs. Two primary approaches to assessing build time include detailed analysis, which uses knowledge about a specific additive manufacturing system, and parametric analysis, which uses information about process time, the volume of material required, layer thickness, and other AM characteristics (Thomas and Gilbert, 2014).

Luckily, various software tools can estimate total print time for an STL or similar file imported to an additive manufacturing system. The software will make calculations and estimations based on a wide variety of parameters and machine settings and provide an estimated build time specific to the part design, material, AM process, and machine used. AM systems may include build estimation software tools specific to the AM system.

AM Notable 7.2

AM for AM: UAM Facilitating PBF In Situ Monitoring

Fabrisonic LLC uses its innovative ultrasonic additive manufacturing (UAM) process to facilitate in situ quality control in powder bed fusion processes (Zelinski, 2020a). The new "smart" build plate produced with UAM contains fiber optic sensing that detects build failure. It monitors real-time data on temperature, stress, and strain while printing a metal part in a powder bed fusion process, selective laser melting (SLM). A grid of sensors across the plate surface gathers strain vectors, and if strain changes abruptly, it indicates a problem.

It doesn't save the part from being scrapped but saves time by preventing the rest of the build from happening rather than discovering the failure after hours of build time.

UAM is a room-temperature, solid-state process that builds solids out of thin layers of metal foil using ultrasound to bond them. SLM requires very high temperatures to melt metal. UAM was the AM process necessary to produce an SLM build plate with sensors that could handle the heat. The sensors, encased between metal sheets during the ultrasound bonding process, are protected from high temperatures.

BUILD PROCESS

The final edited file and selected parameters are software inputs used to translate the design into the slices representing single layers to the 3D printer. The slices, counters, and hatches produced from the STL file make up the build instructions sent to the 3D printer by the build processor. When all machine-specific preparations are complete, the printing process can start.

Monitoring the Build Process

In metal AM, destructive testing or CT scanning can assess quality control and assurance of a part. These approaches are expensive and time-consuming and do not always provide accurate results. Given the need for quality assurance in AM, particularly in higher-risk applications such as aerospace, defense, automotive, and energy applications, other solutions are evolving.

In situ monitoring during the build process is developing rapidly. Monitoring occurs using a variety of sensors as the part is being constructed within the build envelope.

Current technologies used to measure critical parameters during the build process include the following (Wing, I., Gorham, R. and Sniderman, B., no date):

Ultrasound sensors	Examine the part for internal voids
Thermal imaging	Monitors the melt pool relative temperature, size, and shape
Pyrometry	Correlates temperature to the light intensity measured at a single point
High-resolution photography	Facilitates almost real-time inspection of parts in the build chamber
Accelerometers	Measure vibration of the printhead to detect possible anomalies (FDM)

Data generated by the sensors can update the build process iteratively in real time. The ability to control the build process can help manufacturers reach quality goals by facilitating consistent geometries, desired surface finishes, and specified material properties.

Safety Issues

The build process represents the AM stage at which potential occupational health and safety issues need to be identified and mitigated. Due to AM system designs that mitigate most risks to human operators, material handling and equipment environment tend to draw the most attention.

Emission of particulates, including ultrafine particles (UFPs) and volatile organic compounds (VOCs), have been confirmed from FDM and binder jetting processes in both office and laboratory settings. Harsh skin reactions, eye irritation, and allergies have occurred with operator exposure through inhalation or skin contact with some materials used in these processes. Some photopolymer resins and solvents may be harmful and cause symptoms such as respiratory difficulty and skin burns (Buranská et al., 2019). Table 7.2 includes a summary of potential hazards related to AM processes, technologies, and materials.

Education for operators on the handling and disposal of potentially hazardous AM raw materials and handling high-intensity laser beams or similar energy sources will help mitigate potential risks. Standard working equipment should include masks, goggles, and gloves in the AM equipment and work areas. Proper dust collection and air ventilation are also essential for the work area when fine particulates are involved.

Table 7.2 AM Technology and Potential Hazards

AM Technology: Source	AM Material and Form	Process	Potential Hazards
Material extrusion	Thermoplastics, with additives; as filament, granule, pellet, or powder	Electric heating element-induced melting and cooling	• Inhalation exposure to VoCs, particulate, or additives • Skin burns
Powder bed fusion	Metal, ceramic, or plastic; as powder	High-powered laser or electron beam heating	• Inhalation/dermal exposure to powder and fumes • Explosion • Laser/radiation exposure
Vat photopolymerization	Photopolymers; as liquid resin	Ultraviolet-laser-induced curing	• Inhalation of VoCs • Dermal exposure to resins and solvents • Ultraviolet exposure
Material jetting	Photopolymer or wax; as liquid ink	Ultraviolet-light-induced curing	• Inhalation of VoCs • Dermal exposure to liquids and solvents • Ultraviolet exposure
Binder jetting	Metal, ceramic, plastic, or sand; as powder	Adhesion	• Inhalation/dermal exposure to powder • Explosion • Inhalation of VoCs • Dermal exposure to binder/adhesive

(Continued)

Table 7.2 (Continued)

AM Technology: Source	AM Material and Form	Process	Potential Hazards
Directed energy deposition	Metal, ceramic, or plastic; filament or powder	High-powered laser, plasma arc, or electron beam melting	• Inhalation of powder or fumes • Dermal exposure to powder • Skin burns • Laser/radiation exposure
Sheet lamination	Metal, ceramic, or plastic; solids	Rolled film or sheet	• Inhalation of fumes and VoCs • Shock • Laser/radiation exposure

Source: Adapted from (Roth et al., 2019).

AM Notable 7.3

Why Builds Fail, and Now What?

Even given the best planning, builds fail. Let's explore common build failure causes and their solutions:

Design File Errors or Missing Parts

Errors in a CAD design file will transfer to the STL file for printing, and if not corrected, these errors are likely to result in a build failure. For example, rough edges or gaps between surfaces are unlikely to fuse correctly between layers, causing parts to be structurally unsafe and risking the entire build's collapse. Designs should be fully analyzed in CAD and fixed or rebuilt before being sent to the 3D printer for another build attempt. Also, by designing a part as a whole rather than separate pieces, leveraging AM design and manufacturing capabilities, missing pieces are prevented.

Successful Build, but Supports Can't Be Removed

If the initial design did not provide easily removable supports, changing the part orientation on the build platform may eliminate the need for support. Support redesign may be necessary to make supports break away more easily, or if feasible, explore soluble support materials. In a redesign, make sure the maximum support angle is 45 degrees.

Unsupported Part Details

Details designed on parts have tolerances that need to be incorporated, or a build may fail. Details that are too thin or fragile may bend or break during removal from the build chamber or during support removal. Standard tolerances for parts may include embossing, engraving, horizontal and vertical holes, smallest feature possible, pin diameter, and maximum angle of 45 degrees.

Witness Marks Hard to Finish

A part may print successfully, but it may be impossible to complete to required standards. Witness marks are the marks left from support removal. If supports are difficult to remove, witness marks are likely difficult to finish, too. Supports located in hard-to-reach places create these issues. The design process should consider support placement implications for finishing, as well as ease of removal.

Residual Stresses

Residual stresses occur in metal AM and are caused by temperature gradients from the surface to the center of a part during the cooling process. The material cools more quickly on the surface than at the core, which produces residual stresses that cause the part to detach from the build platform. Design parts to reduce residual stress by avoiding large masses of material, having even wall thicknesses, and using small hatching patterns on part surfaces. Process setting and post-print heat treatments can also address residual stress (Diegel and Wholers, 2018).

Case Conclusion: Build and Assess Quality

The GWM AM Pilot team completed their session with Doug and were delighted to observe the "flip of the switch" the powder bed fusion machine fired up, preheated the build plate, and began to print their part.

Their takeaways from this experience include the following:

- A variety of variables impact AM quality, most notably:
  - Design imperfections
  - STL file and build instruction errors
  - AM system build parameters
  - Material quality
  - Energy sources and strengths
  - Designer and operator knowledge
- AM quality can be assured by:
  - Utilizing software that can identify and fix design imperfections and file errors
  - Modeling and simulating the build process using sophisticated software
  - Ensuring operators and material handlers receive training to support material quality
  - Ensuring designers receive training in AM design processes
  - Using sensors to monitor and adjust the build process as it occurs

AM parts are subject to many quality assurance processes applied to traditionally built parts. Although AM systems and processes are relatively safe, you should always observe standard occupational health and safety issues in the workplace.

REFERENCES

3D printing for quality assurance in manufacturing | Deloitte Insights (2015). www2 .deloitte.com/us/en/insights/focus/3d-opportunity/3d-printing-quality-assurance-in-manufacturing.html (accessed 3 September 2020).

Buranská, E. et al. (2019). Environment and safety impacts of additive manufacturing: A review. *Research Papers Faculty of Materials Science and Technology Slovak University of Technology*, 27(44), pp. 9–20. doi: 10.2478/rput-2019-0001.

Diegel, O., and Wholers, T. (2018). How residual stress can cause major build failures, and what you can do to prevent it. *Metal Additive Manufacturing*, 4 (4), 125–128. www .metal-am.com/articles/how-residual-stress-can-cause-major-build-failures-in-3d-printing/ (accessed: 8 November 2020).

Dowling, L. et al. (2020). A review of critical repeatability and reproducibility issues in powder bed fusion. *Materials & Design,* 186, p. 108346. doi: 10.1016/j.matdes.2019 .108346.

Materialise Magics Modules. (2020). Materialise. www.materialise.com/en/software/ magics/modules (accessed 16 September 2020).

Pal, D., Patil, N., Nikoukar, M., Zeng, K., Kutty, K.H., & Stucker, B. (2013). An Integrated Approach to Cyber-Enabled Additive Manufacturing using Physics based , Coupled Multi-scale Process Modeling. https://pdfs.semanticscholar.org/709d/d2d3ec7b0a427d 40559c67d45c40ef23b76a.pdf?_ga=2.100521757.201420326.1607976077-1653444825 .1607976077 (accessed 09 September, 2020).

EOS (2020). Quality assurance in 3D printing. www.eos.info/en/industrial-3d-printing/ 3d-printing-quality-assurance (accessed 9 November 2020).

Pal, D. et al. (2013) An Integrated Approach to Cyber-Enabled Additive Manufacturing using Physics based , Coupled Multi-scale Process Modeling. In Conference: International Solid Freeform Fabrication Symposium, Austin, TX.

Roth, G. A. et al. (2019). Potential occupational hazards of additive manufacturing. *Journal of Occupational and Environmental Hygiene,* 16 (5), 321–328. doi: 10.1080/ 15459624.2019.1591627.

Thomas, D. S., and Gilbert, S. W. (2014). *Costs and Cost Effectiveness of Additive Manufacturing.* NIST SP 1176. National Institute of Standards and Technology, p. NIST SP 1176. doi: 10.6028/NIST.SP.1176.

Wing, I., Gorham, R., and Sniderman, B. (2015). *3D Opportunity for Quality Assurance and Parts Qualification.* Deloitte University Press. www2.deloitte.com/content/dam/ insights/us/articles/3d-printing-quality-assurance-in-manufacturing/DUP_1410 -3D-opportunity-QA_MASTER1.pdf (accessed 07 July 2020).

Zelinski, P. (2020). AM helping AM: 3D printed build plate detects build failures. *Additive Manufacturing* (March 25). www.additivemanufacturing.media/blog/ post/am-helping-am-3d-printed-build-plate-detects-build-failures (accessed 8 November 2020).

Zelinski, P. (2020). Your Metal AM Process Development Tool? A Spreadsheet (Includes Video). *Additive Manufacturing* (August 24). www.additivemanufacturing.media/blog/ post/your-metal-am-process-development-tool-a-spreadsheet-includes-video (accessed 16 September 2020).

Chapter 8

What Happens after the Additive Manufacturing Build Process Is Complete?

Case Introduction: Managing Post-Processing Tasks

The GWM AM Pilot team's newly designed camp stove burner completed within 12 minutes of the estimated print time that Doug worked out with the operator and the team. The group met to review the build operation and plan for post-processing. An excerpt from their conversation follows:

BOB: It's great to see the actual part finally – I feel like we've been working on this project for a long time, and now our dream has taken shape! I can't wait to take this back to GWM and run it through performance testing.

Fundamentals of Additive Manfacturing for the Practitioner, First Edition. Sheku Kamara and Kathy S. Faggiani © 2021 John Wiley & Sons, Inc. Published 2021 by John Wiley & Sons, Inc.

ROXIE: Are we really ready for that? I want to verify the part with a post-build inspection to see how it aligns with our design specifications and mechanical properties before we get to functional testing.

DOUG: Good call, Roxie! We'll need to take care of a few post-processing steps before we're ready to examine the mechanical properties. Since we used powder bed fusion and stainless steel, we'll need to complete a few tasks before the part is prepared to go.

PETE: Ed and I are curious about how much time all this post-processing will add to the overall production time, should the part tests be successful. And we're sure Frank is worried about adding additional cost to the part and camp stove overall.

FRANK: I am concerned about more time and money, as always. Doug, let's get going on the post-processing steps. I'm anxious to learn more!

INTRODUCTION

When 3D printing stops, post-processing begins. Most parts produced by AM require post-production processing to meet the design requirements or functional specifications for their intended application. An analysis of the impact of post-processing on time, cost, and quality revealed average post-processing increased AM production costs by $25,000–50,000 per AM machine per year, while the length of time to finished parts increased by 17–100% (Grimm, T., 2019).

Post-processing typically incorporates tasks to be completed at the end of the build and before the secondary processes applied to some traditionally produced or AM parts. Confusion exists between what constitutes primary vs. secondary post-processing across the AM industry. Grimm (2019) categorizes AM post-processing as primary and secondary and explains them this way:

Primary post-processing includes the steps that need to be performed on all parts to make them suitable for use in any application. The steps vary by technology but generally include cleaning and support structure removal. Experienced users often refer to this basic finishing level as "strip and ship" because the resulting parts lack any embellishment that enhances quality.

Secondary post-processing includes optional part finishing that improves the aesthetics or function of the part. Secondary post-process commonly includes sanding, filling, priming, and painting; however, it can also include machining, plating, or other tasks.

According to the 2020 Additive Manufacturing Post-Printing Survey (PostProcess, 2020), support removal and surface finishing are the most common post-processing tasks performed by operators. They account for over 50% of post-processing time. Though predominately manual, post-processing operations may also be automated or semi-automated and performed in serial or batch processes. Automated post-processes for support removal and surface finishing are becoming more prevalent.

AM post-processing requirements are process, technology, and material dependent, as are design and build activities described in previous chapters. This chapter presents the following primary post-processing steps, along with differences across AM technology and materials:

- *Part Removal:* Platform and part removal
- *Material Removal:* Unused or uncured material cleaning from part
- *Support Removal:* Support removal, if necessary
- *Treating:* Curing, infiltrating, or heating
- *Finishing:* Repairing, joining, or final finishing
- *Inspection:* Testing and inspection for quality assurance

PART REMOVAL

The platform or part removal step occurs when the build process ends. Based on the technology or material used, the operator removes an individual part from the machine or removes the entire build platform before removing the parts. In some cases, the operator removes the entire system comprised of the platform, parts, and material from the machine.

In binder jetting, parts are not attached to the build platform but produced on a powdered bed, so removing the part from the build chamber is simple. The operator vacuums out excess powdered material from the build chamber after the build process. Prior knowledge of the part layout on the platform is critical to ensure that smaller parts are not sucked into the vacuum or damaged during powder removal from the build area. Since the powder is not heated, it is easier to brush it off the parts. Figure 8.1 shows part removal from the build platform.

Figure 8.1 Removing Part from Build Platform
Source: guruXOX/Shutterstock.com.

The binding agent's initial fusing of powdered particles during the binder jetting process produces parts that can be fragile in their "green" state before post-processing. Post-processing treatment is needed for some materials to make the parts fully dense and easier to handle. Due to this fragility, smaller features and complex geometry can break off as parts are removed from the powder bed, requiring care during the removal process.

High-resolution features printed in material jetting and vat photopolymerization systems result in part features easily damaged during post-processing. Fixtures designed to prevent breakage and support part features, like those used in traditional manufacturing, are used until final finishing to avoid damage.

In direct energy deposition, the parts are welded directly onto the build platform or directly on an existing part. In some cases, the platform is part of the printed component, given that supports are not used with this AM process. The operator uses a bandsaw or wire EDM to remove parts from the platform.

Most low-cost material extrusion systems with open architectures do not require significant removal efforts as parts simply scrape from the build platform or base. In large mainframe machines, the operator removes the build platform before removing parts because the enclosed build volume may be at elevated temperatures depending on the material used.

Material jetting requires supports and uses a removable platform. Parts are generally scraped from the platform using a taping knife. Careful handling of parts during removal avoids breakage.

MATERIAL REMOVAL

In most AM processes, uncured, unsintered, or unmelted powder forms a film or coating on the surface of completed parts. The operator must remove this excess or unused material. The techniques used for removal depend on the material and process involved. Most photopolymer-based materials are removed with solvents. Sand or bead blasting removes powdered-based materials. This step is critical to the part's accuracy. The excess material can fuse on the part surface and form an additional layer maintained in subsequent post-processing steps, resulting in inaccurate part dimensions.

The operator can easily brush unbonded powder particles off finished parts since no secondary material except the binding agent fuses the parts. Figure 8.2 illustrates the removal of excess material in binder jetting.

In material extrusion systems with single and multi-extrusion head systems, no residue or unused material accumulates on the part surface. The nozzle extrudes the exact amount of material needed, and the material hardens for every layer.

In material jetting, the jetted material is cured immediately with photopolymer-based systems, and there is no need to remove uncured residue. In wax-based systems, the material hardens before the next layer application during the part build.

Figure 8.2 Removing Excess Material after Binder Jetting Build
Source: Courtesy of MSOE.

Figure 8.3 Removing Unused Material from Parts
Source: Courtesy of Formlabs.

In powder bed fusion for metals and polymers, the parts are bead-blasted to remove the un-sintered powder that adheres to the part surface because of the high thermal energy employed in the process. Bead-blasting is also used by the operator to remove loose powder in cavities and holes since powder bed fusion uses no supports.

Vat photopolymerization systems require solvents to remove any residue of resins from the part surface before further post-processing. Parts produced in this AM process are generally not fully cured. An additional step utilizing a post-cure apparatus is required to fully cure the parts and make them safe for handling without the need for protection. Figure 8.3 illustrates removal of unused material from parts.

In liquid- and powder-based systems, designers and operators must ensure there is no trapped volume within the part. Trapped volume is the calculable volume of the unprocessed material entrapped inside parts with hollow features (Kim *et al.*, 1998). When the trapped volume is significant, the part can be designed with an opening or subjected to additional post-processing to drill a hole for material removal.

SUPPORT REMOVAL

Some AM processes and material combinations require supports. The operator removes the supports from the parts during post-processing before the final curing or hardening of a part occurs, as applicable. There are examples where the support structures are used as fixtures to hold critical features on parts in a specific position due to part geometry and size. In these cases, support removal may be the final step in the process.

Techniques for removing supports include machining, break-away, and using soluble materials. In some AM processes, supports are printed with secondary materials. In these cases, the support can easily be detached or dissolved. Soluble support materials, introduced in the late 1990s, are available for most material extrusion systems. Some high performance and high-temperature materials still use breakaway support materials.

For example, Stratasys' Ultem™ material uses break-away supports that can easily be removed by hand. In contrast, their SR-30 soluble support material used for ABS parts dissolves entirely in a sodium hydroxide solution eliminating manual support removal (Stratasys, 2020). Figure 8.4 shows the removal of soluble supports from a 3D-printed part.

Metal-based powder bed fusion technologies require support structures that can be challenging to remove by hand. Supports and parts are made from the same material, requiring manual machining and grinding for manual support removal. However, more complex part geometries may not lend themselves to machining methods to remove supports (Simpson, 2017).

Renishaw and Additive Automations have partnered to utilize collaborative robots (cobots) with deep learning to remove supports automatically (Renishaw, 2020). Their partnership's focus is to automate support removal, one of the few remaining AM processes that are mostly manual.

Automated systems are available for support removal only or support removal and surface finishing. These systems can remove supports and finish parts from both polymer and metal materials. Due to the repeatability and chemicals used in these processes, the resultant parts generally have superior mechanical properties compared to unfinished parts and those finished manually.

Figure 8.4 Removing Water Soluble Supports
Source: Photo Credit: Stratasys

TREATING

Some combinations of AM processes and materials produce parts that require post-processing to achieve desired mechanical properties or design requirements. Heat treatments, such as sintering, infiltration with other materials, and curing to harden parts are examples of these types of post-processing techniques. Parts in their "green" states after 3D printing may be fragile or pose safety issues due to potential contact with uncured resins.

In binder jetting, metal parts need to be sintered or heat-treated to gain strength. Placing the parts in a high-temperature furnace burns out the binding agent and bonds the powdered metal particles together, reducing porosity and increasing strength. Parts may also require infiltration with other materials, such as a low melting temperature metal like bronze, to enhance mechanical properties. During the furnace cycle, as the binding agent burns out at higher temperatures, it is replaced by the infiltrate to reduce porosity and increase strength (Varotsis, 2020). In plaster-based binder jetting, an infiltrate of cyanoacrylate is added to green parts after the removal of unused powder to give them superior mechanical properties. Material extrusion parts do not require this step.

In vat photopolymerization, the process is material dependent. Some high-temperature materials require both ultraviolet curing in a post-cure apparatus, coupled with cycled heating in an oven to achieve full mechanical properties. For example, Somos® PerFORM's heat deflection temperature increases from 270°F to 514°F after two hours of thermal post-cure heat treatment at 160°F (Somos, no date).

FINISHING

During the finishing step, final techniques are applied by the operator or finishing technician to ensure the realization of desired surface quality, geometric accuracy, and mechanical properties. Finishing techniques may include machining to add holes or other functional features, blasting and polishing to address surface finish requirements, or polishing internal surfaces such as cooling channels to help address performance needs. In some cases, individual sections are subject to sanding, priming, painting, or other processes before joining.

A key task in finishing is to assemble large parts divided into sections to fit the build platform and printed separately. During the design process, the designer and operator work with easily manipulated STL files and standalone software to create unique sections. The end-use application may influence part section placement, the design of sections, and the cut orientation. For example, for design validation models, a simple L- or U-shaped cut may suffice and allows for easy alignment along the X- and Y-axis. A lap-joint or alignment along all axes may be needed to add additional strength for functional models. Depending on the bonding agent's viscosity for joining part sections, operators can account for the tolerance when the sections are created in CAD, resulting in a perfect fit. Figure 8.5 illustrates joining lattice sections of a part during the finishing step.

Figure 8.5 Joining Lattice Components After 3D Printing
Source: Courtesy of Materialise.

INSPECTION

Inspection processes for post-production quality control and evaluation of AM parts are similar to those for traditionally manufactured parts and may include these steps:

- Measure the part dimensions to ensure the part is within all tolerances.
- Visually inspect the surface finishes for intact features and aesthetics, as defined by the application; AM parts are inspected at part removal and again after post-processing.
- Check for flatness or warp of the part.
- Identify debris or foreign objects incorporated into the part.

With 3D-printed parts subject to redesign and consolidation, nondestructive inspection techniques that use CT scanning are ideal for evaluating internal geometries.

SUMMARY OF POST-PROCESSING METHODS

In a recent study, 75% of respondents indicated they perform three or more post-processing steps to get to their desired final part (PostProcess, 2020). They also reported that finding skilled labor to conduct post-processing activities at consistent quality levels is an increasing concern. It is no wonder that as companies move to take AM to production scale, post-processing is receiving greater attention as a predominately manual task with the highest post-printing costs in the overall AM process flow. New adopters of AM should consider automated options for post-processing, since post-processing poses a challenge to moving to scale with AM. Table 8.1 summarizes the post-processes commonly used with AM technologies and materials.

Table 8.1 Common Post-Processing for AM Technology and Materials

AM Process/ Post-Process	Part Removal	Material Removal	Support Removal	Treating	Finishing	Inspection
Metals: Powder bed fusion	Y	Y		Y	Y	Y
Metals: Binder jetting	Y	Y		Y	Y (optional)	Y
Metals: Direct energy deposition	Y			Y	Y	Y
Plastics: Material extrusion			Y		Y	Y
Plastics: Vat photopolymerization		Y	Y	Y	Y	Y
Plastics: Powder bed fusion		Y			Y	Y
Plastics: Material jetting			Y			Y
Plastics: Binder jetting		Y		Y		Y

Source: Adapted from AMFG, 2018.

Case Conclusion: Taking Post-Processing to Scale

After completing the necessary post-processing steps, the camp stove burner was ready to return to GWM for functional and performance testing. The part passed all the required tests and hit all benchmarks. The GWM AM Pilot team met to review the results and prepare their recommendation for GWM's leadership group. Here is part of the conversation:

BOB: I think our customers will appreciate that they no longer need to perform assembly on the burner component – and so will our customer support folks! Plus, the reduction in the stove's overall weight resulting from the redesign should be something our marketing team emphasizes when they roll out the redesigned product.

FRANK: I'm amazed that we can reduce the amount of material that goes into the new burner. That's going to be significant cost savings, even with the additional post-processing steps.

ROXIE: It looks like we can build enough monitoring into the AM production process to meet the requirements for our quality assurance program and to meet our product safety standards. I'm looking forward to other applications we can explore with AM!

PETE: We're talking like this is a done deal – hope we can convince leadership that moving into AM is the right way to go! I'll need to get together with HR and talk about how we reskill our production employees and hire a few new AM specialists.

ED: Okay, guys! Is it finally time to celebrate our successful pilot with AM?

REFERENCES

AMFG. (2018). Post-processing for industrial 3D printing: The road towards automation. *AMFG*. amfg.ai/2018/12/11/post-processing-industrial-3d-printing-road-towards-automation/ (accessed: 26 October 2020).

Grimm, T. (2019). *3D Printing: The Impact of Post-Processing – Tech Briefs*. www.techbriefs.com/component/content/article/tb/pub/features/articles/33589 (accessed: 25 October 2020).

Kim, J. Y. et al. (1998). Efficient calculation of trapped volumes in the layered manufacturing process, *The International Journal of Advanced Manufacturing Technology*, 14 (12), 882–888. doi: 10.1007/BF01179077.

PostProcess. (2020). *Annual Additive Post-Printing Survey: Trends Report 2019*, www.postprocess.com/trends-2020/, p. 14.

Renishaw (2020). *Renishaw joins project to automate additive manufacturing post-processing*. www.renishaw.com/en/renishaw-joins-project-to-automate-additive-manufacturing-post-processing--45601 (accessed: 26 October 2020).

Simpson, T. (2017). *Removing metal supports from AM parts.* www.additivemanufacturing
.media/blog/post/removing-metal-supports-from-am-parts(2) (accessed: 26
October 2020).

Somos (no date). Somos perFORM. www.dsm.com/content/dam/dsm/somos/en_US
/documents/Brand-Status-Sell-Sheets/English-Letter/Somos%20PerFORM%20SS-PDS
%20Letter.pdf (accessed: 26 October 2020).

Stratasys (2020). *Stratasys FDM Support Materials | Stratasys™ Support Center.* support
.stratasys.com/en/materials/fdm-materials/fdm-support-materials (accessed: 26
October 2020).

Varotsis, A. (2020). *Introduction to binder jetting 3D printing, 3D Hubs.* www.3dhubs.com
/knowledge-base/introduction-binder-jetting-3d-printing/ (accessed: 26 October 2020).

Chapter 9

What Do We Do to Move to Additive Manufacturing?

INTRODUCTION

The results of a recent independent global study reported by *Industry Week* (2020) show that acceptance of additive manufacturing is growing across all industries. 3D printing for full-scale production parts doubled in one year from 2018 to 2019, with continued growth expected (Fretty, 2020). It won't be long before competitive pressures, environmental regulation, and customer requirements begin to exert additional pressure on manufacturers to make a move to AM.

Great West Manufacturing (GWM), the organization described in the case study excerpts included in Chapters 2 through 8, undertook its first steps toward AM. GWM formed a pilot project team, identified a part to experiment with, and walked through the end-to-end process of producing the part with AM. Now that the pilot project is complete and achieved positive results, GWM must make some important decisions to move forward with AM. This chapter sets out a roadmap for the next steps in the AM adoption process by providing a summary of the AM implementation process and highlighting key

Fundamentals of Additive Manfacturing for the Practitioner, First Edition. Sheku Kamara and Kathy S. Faggiani © 2021 John Wiley & Sons, Inc. Published 2021 by John Wiley & Sons, Inc.

decisions and factors to address to ensure implementation success. Any organization can benefit from careful consideration of the roadmap provided here.

AM IMPLEMENTATION ROADMAP

Manufacturing firms are notorious for being slow to change and risk-averse, which is understandable given the level of investment in existing infrastructure, costs to retool, and the magnitude of change required across the entire organization to address major technical advancements that drive growth. The roadmap depicted in Figure 9.1 is designed to foster thinking about moving to AM successfully.

GWM began their journey to AM with a pilot test broadly communicated throughout the organization. The well-communicated pilot allowed the organization to learn more about the technology while considering the best applications within their existing operation. While conducting a pilot project is a good approach to exploring AM capabilities, other ways of initiating an investigation are possible. Consider Heineken's approach, described in Chapter 2, which involved acquiring an AM system and installing it on the plant floor to allow workers to identify applications. Other strategies include consulting with a local university AM consortium, service bureau, or reputable AM systems vendor to gain insights and develop an exploration path.

After vetting AM as a manufacturing approach and deciding to move forward, the next critical step is developing a vision and strategy for AM as a part of the business model. Depending on how the organization will implement AM, leaders must prepare a change management plan to ensure implementation success. Beyond the need to acquire AM system technologies and support, the organization's AM workers will need training. New processes and workflows will need to be defined and implemented to address AM's

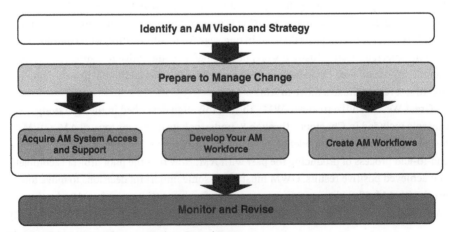

Figure 9.1 AM Implementation Roadmap
Source: Adapted from Mellor, S. 2014.

integration with manufacturing operations. Finally, as the implementation progresses, key indicators of impact should be measured and monitored to ensure AM vision and strategy realization. The following sections describe each of the implementation roadmap steps in more depth.

IDENTIFY A VISION AND STRATEGY

Dr. Scholl's, a 113-year-old brand and leader in foot care, recognized the importance of customization to address consumer needs. A partnership with Wiivv Fit Technology moved the company quickly into AM. It scaled up its ability to provide shoe inserts that provided the right support in the right places based on the exact specifications of a customer's feet (Scott, 2019). Wiivv Fit and Dr. Scholl's custom 3D printed insert is generated from images produced using a simple smartphone app that maps 400 points on a customer's foot, allowing customers to order a fully customized product from home. Profitable mass customization, enabled by AM, was aligned with Wiiv and Dr. Scholl's business philosophy to provide customized foot care for consumers.

As discussed in Chapters 2 and 3, a wide variety of ways to apply AM exist. The key to developing a vision and strategy for successfully adopting AM rests in selecting an application aligned with the business, manufacturing, research, and development goals and activities. Table 9.1 summarizes different ways to move to AM, potential applications, and their advantages.

An essential element of the AM vision and strategy should be to develop metrics or key performance indicators designed to measure AM goal progress and achievement. Appropriate AM metrics will be dependent on the particulars of a vision and strategy within each company and may involve existing manufacturing KPIs already in use.

For example, Table 9.1 references "excessive downtime from machine breakdowns" as a problem to be addressed using AM. A company adopting AM to generate replacement parts might choose the following metrics to monitor progress:

- Time from breakdown to part replacement
- Percent of machine downtime due to breakdown

Another example from Table 9.1 is "High degree of assembly time and cost on products." By implementing an AM system and using design for AM advantages, the company could select metrics from the following depending on specific circumstances:

- Reduction in # of parts required
- Reduction in the labor cost to assemble
- Reduction in time to assemble
- Reduction in overall product cost

Ideally, key performance indicators or metrics should be identified before AM implementation to allow pre- and post-implementation measurements and comparisons.

Table 9.1 Ways to Move to AM

For this situation. . .	Consider using AM for. . .	Potential advantages
Long concept-to-production cycle	Rapid prototyping, proof of concept, functional testing, final design approval, limited production	Reduce product development cycle time Reduce time to market
Excessive downtime from machine breakdowns	Producing replacement parts	Reduce time to acquire parts to repair machinery, thus reducing downtime
Costly and time-consuming creation of jigs and fixtures	3D printing tooling for the shop floor	Reduce costs and time for tooling in traditional manufacturing processes
Excessive cost/lack of availability of spare parts Legacy parts/low volume (single part count)	Producing replacement parts, no tooling costs for single part replacement	Produce replacement parts in house to save time and money; particularly valuable when parts are hard to find or must be shipped internationally
Unmet customer demand for customization	Customizing products	Provide cost-effective customization of products without need for retooling and major production changes
High part complexity, too costly or time consuming	Redesign and production of existing parts to reduce complexity	Redesign existing parts to reduce complexity and save time and money
High degree of assembly time and cost on product	Redesign and production of existing parts to reduce number of parts and associated assembly time and cost of product	Redesign product to reduce number of parts and minimize need for assembly or make assembly less complicated
Opportunity to innovate/create new products	Design, test, and produce new, innovative products	Create items currently not possible with traditional manufacturing, or customize items not previously customizable
Implement startup company ideas	Producing and customizing products	Imagine and implement product creation and delivery to consumers; enable cost-effective customization where it was not possible before

Source: Original, Kamara and Faggiani, 2020.

PREPARE TO MANAGE CHANGE

AM is a disruptive technology that challenges much of the understanding, practice, and experience of traditional manufacturing. Linde, an industrial gas and engineering company, conducted a research project to explore industrial burner development using AM (Neuner and Lang, 2019). Some of the key challenges encountered were human rather than strategic or technical failures. Specifically, developing AM knowledge within manufacturing engineering teams and overcoming resistance to AM innovative technologies proved difficult. Linde produced a successful AM prototype of a new industrial gas burner through trial and error, which helped engineers learn AM design capabilities. Figure 9.2 shows the Linde burner.

Linde's experience highlights the change in mindset required to successfully implement AM, not only on the part of designers but also on everyone in the organization. Some manufacturing leaders have referred to the need to change organizational culture to foster a less risk-averse and innovative workplace that will embrace AM. In contrast, others claim they will delay AM adoption until the technologies and processes are more mature and widespread. At present, industries have implemented the changes required for AM at different paces . The medical sector and aerospace currently lead the pack, and the construction industry is just beginning to explore AM possibilities.

Design, the most crucial area of 3D printing, also needs the most change to take full advantage of AM capabilities as learned at Linde. In a 2020 study (Langnau, 2020), more than half of participating companies reported the need to develop design expertise to scale AM production. Other key organizational changes needed include developing new procurement strategies and processes, changing production team mindset, and developing trusted total cost of ownership (TCO) and return on investment (ROI) models.

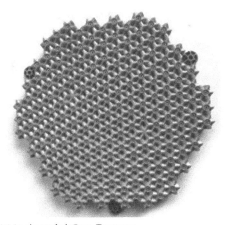

Figure 9.2 Linde AM Industrial Gas Burner
Source: Image provided by courtesy of Linde GmbH, Germany, 2020.

Addressing the scope and magnitude of change involved in taking AM to scale in a manufacturing operation requires careful attention and planning. The necessary changes must be managed and guided to ensure success. No "one size fits all" change management plan is possible since the range of changes will vary depending on the AM application selected, organization size, and AM technology employed. However, three areas are relevant to almost all AM adoptions: acquisition of AM systems and support, development of AM workforce knowledge and skills, and redesigning or creating processes and workflows needed to support the integration of AM into the organization.

ACQUIRE AM SYSTEMS ACCESS AND SUPPORT

A variety of options for gaining access to AM systems and support are available to companies moving into AM. In general, sourcing AM system access should consider the AM vision and strategy, the intended AM application(s), and each application's business case. Figure 9.3 shows options that include purchasing or leasing an in-house AM system, joining a university AM consortium, contracting with an AM service bureau, or collaborating with cross-industry partners.

Consider In-House AM Adoption

Purchasing and implementing an in-house AM system(s) for serial production is likely the costliest path forward but may yield the most significant returns. Thomas and Gilbert

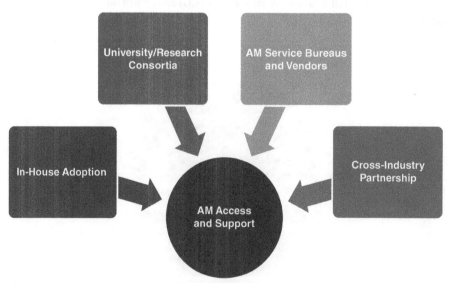

Figure 9.3 AM Access and Support Options
Source: Original, Kamara and Faggiani, 2020.

(2014) analyzed the costs and cost-effectiveness of AM and identified four key cost areas in AM implementation that represent the total cost of part production. Figure 9.4 shows the most considerable portion of the total cost is initial machine cost, while the second-largest AM production cost is materials. Labor and post-processing costs round out the total cost profile. It is important to note that each category's costs will vary by AM technology and material used. The cost figures presented represent broad averages.

Additional costs from ongoing system maintenance, materials procurement, and training/retraining employees are also likely to be realized. If these costs and expenses are justified by anticipated production volume and use frequency, an in-house system makes sense.

The decision to adopt AM through the adoption of an in-house system is likely to involve the following additional costs and issues:

- Significant upfront investment
- Facilities changes for additional space and efficient layout of operations, adequate power supply, enhanced environment control/HVAC systems, material storage, and safety assurance
- Increased energy/power utilization
- Post-processing equipment, set up, and ongoing costs
- Hiring or training personnel for AM design and operations
- Establishing new supply chains and procurement strategies
- Developing new production processes and workflows

Suppose the primary applications for AM are creating scale models or functional prototypes that may involve multiple iterations. In this case, an in-house AM system may be the way to go. New and affordable desktop systems can be purchased and placed in office settings and can support an ongoing stream of parts and prototypes.

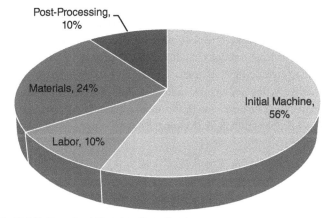

Figure 9.4 AM Estimated Production Costs
Source: Derived from Thomas and Gilbert, 2014.

The benefits likely to accrue from in-house AM systems include protecting intellectual property, producing new ideas internally, on-demand production with quick turnaround and fast iterations, eliminating shipping and delivery charges compared to service bureau use, and relatively lower print costs over time.

AM University Consortia

For small or medium-sized manufacturing companies, buying and installing an AM system may not be a financially feasible option. By establishing a relationship with a university or research institution, these companies can obtain access to AM experts and leading technologies to move toward AM, like GWM's design support, as described in Chapter 4. Some universities have established industry consortia requiring industry members to pay an annual membership fee to cover AM system access, research, and training costs. In contrast, others operate university centers of excellence where industry partners use a pay-as-you-go system.

Proto Precision Manufacturing Solutions, a US metal fabricator, is a partner in the Ohio State University's Center for Design and Manufacturing Excellence (AMFG, 2019). Through the partnership, Proto Precision was able to gain the knowledge and technology support it needed to grow its metal 3D printing capabilities.

A list of AM university consortia and centers of excellence appears in Chapter 10.

AM Service Bureaus and Vendors

Both AM system vendors and service bureaus possess expertise that can help you plan your move to AM. Vendors may provide the option to lease equipment to reduce the amount of upfront investment required to move into 3D printing. By outsourcing AM production to service bureaus when the volume will be low or intermittent may realize considerable cost savings.

Additional benefits can arise from working with a service bureau since they typically offer multiple technologies and materials, allowing more innovation and experimentation with different materials and processes. Also, service bureaus are likely to have a high level of expertise across a broader range of AM issues to assist companies with design and post-processing functions in addition to 3D printing.

The downside to contracting with a service bureau is likely to be slower production times and delivery, depending on location. As Great West Manufacturing (GWM) realized, the nearest service bureau with the metal capability was hours away and may increase costs and delivery time with shipping. Chapter 10 lists selected AM service bureaus and vendors.

Cross-Industry Partnership

Different types of cross-industry partnerships are possible to provide access to AM capabilities. In the Wiivv and Dr. Scholl's example described earlier in the chapter, Wiivv Fit Technology partnered with Dr. Scholl's to help produce custom shoe inserts unique to each customer. Dr. Scholl's needed access to AM technology and expertise to scale up their ability to provide customized products, which Wiivv Fit provided. Also, Wiivv provided the customer-facing web portal for smartphones. This collaborative partnership enabled Dr. Scholl's to move to market quickly with its customized offering.

Another option is to ally with a noncompeting manufacturing firm to purchase excess capacity. The goal is to seek a manufacturing firm in a non-competitive market sector that produces products using the needed AM technology and materials. Suppose another firm has excess AM production capacity available on their machine(s). The firm may be amenable to the purchase of this capacity and possibly expertise. Without the need for retooling, the ability to purchase excess capacity from another manufacturer can be relatively straight forward. Trade organizations and regional AM user groups are useful contacts for identifying companies with AM equipment.

DEVELOP AM WORKFLOWS

Adopting a desktop AM system for rapid prototyping or limited part production involves fewer concerns about process change and workflow than implementing an AM mainframe system for serial or full production. In these situations, the AM desktop systems are likely replacing other limited operations, while the mainframe systems portend a major shift in overall operations.

Incorporating desktop systems may be as simple as adding the prototyping step in the concept-to-design approval sequence or replacing an AM service bureau contract. Modeling the workflows and related processes, estimating build times, and forecasting overall time from start to end of the sub-process involving AM use will help ensure efficient AM use and set AM performance expectations.

Implementing a mainframe system for serial and full production should involve the entire process from initial concept through production, consumption, and final recyclability or disposal of a part or product. Specifically, processes involving design, process planning, quality assurance and control, accounting and financial systems, and integration with or replacement of the existing shop floor will be subject to revision in a broad implementation. AM consultants, vendor representatives, process and systems analysts, and companies with prior experience in AM implementation are useful resources confronting these change challenges.

As AM moves into mainstream manufacturing, demand for a full automation suite of software that integrates all AM processes from order or design through finishing are growing. Given the complexity of workflows, this software is increasingly important to ensure that the right parts are available at the right times in the right location. Allowing engineers and operators to tap into the workflow at any point and see parts awaiting design, printing, post-processing, or finishing can assist in efficient production scheduling and timely delivery.

Bowman Additive Production, a manufacturer of end-use bearings and components, needed to expand its AM production capacity to meet the rising demand for their products, which are not producible using traditional manufacturing methods. Managers quickly discovered they needed an efficient means of handling high production volumes and managing their AM facility.

Bowman's head of production searched for software to manage the production workflow and integrate it with its internal ERP system. More specifically, the goal was to automate the AM process from end to end, including optimizing digital files for printing, scheduling project requests, facilitating efficient machine scheduling, and automatically generating key performance indicator reports. Bowman selected AMFG's AM automation software because it met all of Bowman's needs and enabled it to track AM parts as they moved along the production process (AMFG, 2019).

Workflow software is a valuable tool for scaling AM for full production, according to many industry experts. In addition to providing tracking, reporting, and monitoring of parts as they move through the AM production process, the software needs to interface seamlessly with core systems already in place. Table 9.2 lists a summary of popular AM workflow software packages.

DEVELOP YOUR AM WORKFORCE

Chapter 1 introduced the need for a highly skilled AM workforce and described the knowledge and skills needed by manufacturing employers. The two primary pathways for creating a workforce to move to AM are: (1) train existing employees in AM methods and processes, or (2) hire AM-qualified personnel (contract or full-time) to handle AM processes or train existing workers.

AM Training

AM has matured to the point where two- and four-year college systems offer AM education through degree programs or continuing education offerings. Professional organizations such as SME (SME, 2020) and AMUG (AMUG, 2020) offer certification of individuals who complete training and pass a certifying exam.

AmericaMakes, a public–private partnership working to accelerate AM and global manufacturing competitiveness, has developed a workforce and education roadmap (AmericaMakes, 2020). They offer an apprenticeship program providing manufacturers with training-related services, including the Advanced Curriculum in Additive Design, Engineering, and Manufacturing Innovation (ACADEMI) portfolio of DfAM-based courses. The courses lead to industry certification, and a 3D Veterans program provides a boot camp for 3D printing, software, and design knowledge for veterans with workforce-ready skills.

AM training offered through online delivery typically provides hands-on experience through vendors with actual AM systems operations as part of the system installation and implementation process. When planning the move to AM, managers should include adequate time and cost for employee AM training into the expected startup expenses. Some AM service bureaus have also branched out into offering AM training to clients and others.

Table 9.2 Selected AM Workflow Software

Software/Company	Description	Key characteristics
Link3D Additive MES	Connects upstream to Autodesk Fusion 360 design software and downstream to Autodesk Netfabb	Fully integrated additive manufacturing work-flow software
3D Trust MES	Optimizes workflow from order intake to delivery, real-time data collection, and post-processing; also supports material management	Available on tablet and smartphone
Agile Manufac-turing Software Suite/3YOURMIND	Standardizes and automates full AM process chain	Incorporates Agile Product Lifecycle Management (PLM), Agile Enterprise Resource Planning (ERP), and Agile Manufacturing Execution Systems (MES)
AMFG	Customized product management platform tailored to the needs of each company	Integrates with exist-ing PLM, ERP, and CAD software
Authentise MES	Includes order taking/subcon-tracting, production planning, online 3D model library, daily reporting, and real-time machine monitoring	Designed for OEMs and ser-vice bureaus
Materialise Streamics	Connects AM technologies and digital systems to integrate and streamline AM production processes from request to post-processing and reporting	Supports service bureaus or serial production
Oqton's FactoryOS	Enterprise-wide software that links data from design through production	Startup, with specialized ver-ticals in medical, industrial, and aerospace 3D printing

Source: Adapted and extended from Carlotta V. 2020 and "There Will Be No Industry 4.0 Without AM Workflow Automation," 2018.

AM Hiring

AM hiring is competitive and can be problematic. Metal AM is experiencing rapid growth as more experienced AM manufacturers move from polymer prototyping to serial production. Demand for metal AM professionals is more than double the demand for polymer AM professionals, and difficulties exist in finding candidates with the right skills (Damgaard and Signe, 2020). As of late 2020, there are very few available candidates with a background in metal AM. Many applicants have academic knowledge but little practical or hands-on experience. Given the significant differences between AM processes in polymer vs. metal parts, it is difficult for AM professionals to transition from polymer to metal work roles quickly enough to meet demand. It is a smoother transition to move professionals in traditional metal production to AM metal production, as they are more familiar with metal material capabilities and uses.

Manufacturers should consider offering summer or part-time internships to college students enrolled in programs with an in-depth metal AM component. By training students in practical, hands-on metal AM processes, they can produce a well-qualified future worker in tune with their specific production process. The AmericaMakes Apprenticeship program mentioned in the AM training section above might also provide a means to acquire training-related resources to help polymer professionals leap to metal AM. Focused on-the-job training for those transitioning from polymer to metal AM may provide the fastest pathway to meet metal AM demand. Coupling on-the-job learning with continuing education or professional association education opportunities can both enhance the existing workforce and train new hires.

In addition to the hard skills and knowledge of metal and process engineering, it's worth noting the desire for AM metal hires to possess soft skills that are valuable in AM firms (Damgaard and Signe, 2020). AM metal professionals job candidates should be highly motivated team players, open to new problems and challenges. They should also be curious, ambitious, and able to think creatively or outside the box.

MONITOR RESULTS AND ADJUST

As discussed earlier in this chapter, equally important to developing a vision and strategy for the move to AM is identifying measures or key performance indicators that help assess progress moving toward AM and AM performance within the manufacturing process. An important decision is determining when and where to measure during the process, since timing during the AM implementation could cause misleading results. For example, if the build failure rate is a measure for AM design quality and prebuild activity success, the first builds' measures may not be the right choice. Some degree of experimentation and testing should be factored into the learning process and not reflected in initial measures. Perhaps agreeing to measure within two months or 20 builds may provide a more accurate result of how progress toward the AM vision is faring.

Implementation of AM isn't a "one and done" proposition. Monitoring progress through regular reviews of appropriate measures can help identify needed course corrections before unnecessary costs occur. Manufacturers may wish to consider implementing AM in conjunction with, or with guidance from, Lean manufacturing teams. Many aspects of AM work in parallel with Lean and Six Sigma approaches, the latter of which could provide a supportive change framework around the implementation of AM while ensuring AM is well-integrated into current operations. The discipline and measurement requirements within Lean practices make a good partner to AM implementation.

REFERENCES

AmericaMakes (2020). Workforce & Education. www.americamakes.us/workforce-education/ (accessed: 12 October 2020).

AMFG (2019). 5 ways your company can get started with additive manufacturing [Infographic]. *AMFG* (7 February). amfg.ai/2019/02/07/5-ways-your-company-can-get-started-with-additive-manufacturing-infographic/ (accessed: 6 October 2020).

AMUG (2020). Training & certification. *Additive Manufacturing Users Group*. www.amug.com/2020_train_certify/ (accessed: 12 October 2020).

Carlotta V. (2020). Top additive manufacturing workflow software solutions, 3Dnatives. www.3dnatives.com/en/additive-manufacturing-workflow-software-solutions-300620204/ (accessed: 12 October 2020).

Damgaard, S. (2020) Metal additive manufacturing is on the rise; how and where do you find the right people? *3D Printing Media Network* (October 7). www.3dprintingmedia.network/metal-additive-manufacturing-alexander-daniels-global/ (accessed: 12 October 2020).

Fretty, P. (2020). Survey says: additive manufacturing on the rise. *IndustryWeek* (January 15). www.industryweek.com/technology-and-iiot/article/21120529/survey-says-additive-manufacturing-on-the-rise (accessed 30 September 2020).

Langnau, L. (2020) Essentium research shows change management a priority for additive manufacturing. *Make Parts Fast* (March 20). www.makepartsfast.com/essentium-research-shows-change-management-a-priority-for-additive-manufacturing/ (accessed: 5 October 2020).

Mellor, S., Hao, L., and Zhang, D. (2014). Additive manufacturing: A framework for implementation. *International Journal of Production Economics*, 149, 194–201. doi: 10.1016/j.ijpe.2013.07.008.

Neuner, F., and Lang, F. (2019). Adopting Additive Manufacturing: as much a mindset change as technological. *TCT Magazine* (July 12). www.tctmagazine.com/api/content/a6a111e8-a4a6-11e9-af39-12f1225286c6/ (accessed: 5 October 2020).

Scott, C. (2019). Dr. Scholl's partners with Wiivv for 3D custom insoles. 3D Print.com (January 8). https://3dprint.com/233413/dr-scholls-partners-with-wiivv/ (accessed 11 December 2020).

SME (2020) *Additive Manufacturing Certification Process*. www.sme.org/training/additive-manufacturing-certification/additive-manufacturing-certification-process/ (accessed 12 October 2020).

Thomas, D. S., and Gilbert, S. W. (2014). *Cost and Cost Effectiveness of Additive Manufacturing*. NIST Special Publication 1176. US Department of Commerce, pp. 1–89.

Chapter 10

Where Can I Learn More?

INTRODUCTION

AM continues to evolve rapidly, as do the resources designed to inform, educate, and support manufacturing professionals in moving from traditional methods to additive technologies and approaches. The resources included in this chapter do not represent an exhaustive list of the AM universe. Instead, they have been thoughtfully selected as a set of "good starting points" to explore AM further.

AM ACRONYMS AND TERMINOLOGY

ISO/ASTM52900 (en) Additive manufacturing – General principles – Terminology provides standards established and defined terms in additive manufacturing technology by the International Organization for Standardization (ISO); available for purchase.

www.iso.org/obp/ui/#iso:std:iso-astm:52900:dis:ed-2:v1:en

Additive Manufacturing Glossary – SME provides a list of basic terminology and summaries of key additive manufacturing processes.

www.sme.org/technologies/additive-manufacturing-glossary/

3D Printing Definitions and Acronyms Dictionary contains a list of acronyms and definitions for 3D printing technologies, materials, and processes.

3dsourced.com/guides/3d-printing-definitions-acronyms/

Fundamentals of Additive Manufacturing for the Practitioner, First Edition. Sheku Kamara and Kathy S. Faggiani © 2021 John Wiley & Sons, Inc. Published 2021 by John Wiley & Sons, Inc.

AM Glossary: Terms and Abbreviations contains a wide variety of acronyms and terms used in and around additive manufacturing and also clarifies multiple terms used to refer to the same item.

www.hanser-elibrary.com/doi/pdf/10.3139/9783446431621.bm

AM JOB POSTINGS AND EMPLOYMENT INFORMATION

The following websites list AM jobs, which can be easily accessed using search terms "additive manufacturing" or "3D printing."

General Employment Websites with AM Job Listings:

Indeed.com

www.indeed.com/q-Additive-Manufacturing-jobs.html

Ziprecruiter.com

www.ziprecruiter.com/Jobs/Additive-Manufacturing

SimplyHired

www.simplyhired.com/

Employment Websites Featuring Manufacturing Jobs:

iHireManufacturing

www.ihiremanufacturing.com/

ManufacturingJobs.com

www.manufacturingjobs.com/

Websites Featuring Industry-Specific or Company AM Employment Opportunities:

GE Additive Careers

www.ge.com/additive/careers

Northrup Grumman Manufacturing Careers

www.northropgrumman.com/careers/manufacturing-careers/

AM EDUCATION AND TRAINING

A selection of education and training opportunities in AM is listed below. More two- and four-year universities are beginning to offer degree and non-degree programs to address the AM knowledge gap, so be sure to explore opportunities at local community colleges, technical schools, and universities.

ASME Learning and Development – Additive Manufacturing

ASME offers several AM courses available on-demand and online via a virtual class-room. Courses include: AM with metals, AM part failure analysis, and other topics.

additive.asme.org/#New-Offerings

UL Additive Manufacturing Training and Education

Offers comprehensive AM training programs that lead to Tooling U-SME's professional AM certification.

www.ul.com/resources/additive-manufacturing-training-and-education

MIT Professional Education in Additive Manufacturing

Offers courses that can apply to broader professional certificate programs in technology, design, or manufacturing.

professional.mit.edu/course-catalog/additive-manufacturing-3d-printing-factory-floor

Texas A&M Additive Manufacturing Certificate Program

Certificate program developed and offered as a partnership between the university and AM equipment and materials vendor EOS.

tees.tamu.edu/workforce-development/professional-education/additive-manufacturing-cert/index.html

Penn State Additive Manufacturing and Design

Penn State offers focused graduate degree programs and a graduate certificate in AM.

www.amdprogram.psu.edu/

Tooling U-SME Industry 4.0 Online Training Classes

SME and Tooling-U offer a variety of AM online courses. Just visit their course catalog website and select "Industry 4.0" in the Functional Area box to browse the list.

www.toolingu.com/catalog

AM PROFESSIONAL CERTIFICATION

Completing education or training program requirements may earn a professional certificate from the organization or institution offering the training. Still, industry-wide professional certification – recognized across organizations and industries – is limited for AM. Current options available are:

Additive Manufacturing Certification – SME

SME offers a variety of industry certifications based on job role or focus. Current certifications include:

CAM-F – Certified Additive Manufacturing – Fundamentals

CAM-T – Certified Additive Manufacturing – Technician

SME also offers review courses, training, and certification examinations.

www.sme.org/training/additive-manufacturing-certification/

Stratasys Additive Manufacturing Certification Program

Stratasys offers a program that supplies educational institutions' expertise and resources to encourage student learning in additive manufacturing. At schools that implement the Stratasys program, students can be certified in professional AM technologies and prove their proficiencies in AM applications in various industries.

www.stratasys.com/education/edu-certification

AM BODY OF KNOWLEDGE (AMBOK)

The AMBOK was initially developed in 2014, then updated in 2016. Access it at:

www.sme.org/globalassets/sme.org/training/certifications/additive-manufacturing-certification/additive-bok-2019.pdf

AM UNIVERSITY CONSORTIA AND CENTERS OF EXCELLENCE

AM university consortia and centers of excellence can provide AM expertise, training, research, and information about new developments and AM best practice. A sample of the many groups includes:

National Center for Additive Manufacturing Excellence (NCAME), Auburn University

> http://www.eng.auburn.edu/research/centers/additive/index.html

Golisano Institute for Sustainability, Center of Excellence in Advanced and Sustainable Manufacturing, Rochester Institute of Technology

> www.rit.edu/gis/coe/

Rock Island Arsenal Joint Manufacturing and Technology Center (RIA-JMTC)

> ria-jmtc.ria.army.mil/

ASTM International Additive Manufacturing Center of Excellence

> amcoe.org/

Center for Additive and Digital Advanced Production Technologies, MIT

> http://apt.mit.edu/

Additive Manufacturing Institute of Science & Technology, University of Louisville

> louisville.edu/amist/

Modeling & Optimization Simulation Tools for Additive Manufacturing (MOST-AM), University of Pittsburgh

> www.engineering.pitt.edu/mostam/

W.M. Keck Center for 3D Innovation

> keck.utep.edu/

Rapid Prototyping Center, Milwaukee School of Engineering

> www.msoe.edu/academics/how-we-teach/labs-and-research/engineering/rapid-prototyping-center/

AM SERVICE BUREAUS AND VENDORS
3D Printing Business Directory

This online business directory of all things 3D printing and additive manufacturing includes over 5,300 3D printing companies and is compiled and maintained by the 3D

Printing Media Network. Use this site to perform searches for hardware/equipment, software, materials, service bureaus, post-processing services, and many other categories related to 3D and AM.

www.3dprintingbusiness.directory/

Top United States Additive Manufacturing (3D Printing) Services Suppliers

www.thomasnet.com/articles/top-suppliers/3d-printing-services-companies/

Map Visualizes AM Service Bureaus Across the US

www.additivemanufacturing.media/blog/post/
map-visualizes-service-bureaus-across-the-us

Additive Manufacturing Resource Directory

additivemanufacturing.com/am-service-provider/

AM MATERIAL SELECTION GUIDES

A great source of information about available materials and their properties, including material selection tools and questionnaires designed to assist in the identification of AM materials based on a set of mechanical properties, color, application, certifications, and other factors, are available at the following:

Senvol: http://senvol.com/material-search/

SpecialChem: omnexus.specialchem.com/selectors

Stratasys: www.stratasys.com/materials/search

AM TECHNOLOGY SELECTION GUIDES

Senvol provides a machine search.

http://senvol.com/machine-search/

Selecting the Right 3D Printing Process

Describes how to select an AM process when a material is selected, characteristics of the end part are defined, or when certain process capabilities are necessary.

www.3dhubs.com/knowledge-base/selecting-right-3d-printing-process/

Additive Manufacturing System Selection

Information on AM technology made available by ARC Advisory Group.

www.arcweb.com/technology-evaluation-and-selection/additive-manufacturing-system-selection

AM DESIGN GUIDELINES

Materialise, an AM equipment vendor, provides design guidelines for metals and plastics and several different AM processes and technologies.

www.materialise.com/en/academy-manufacturing/resources/design-guidelines

Design Guidelines for 3D Printing & Additive Manufacturing

The application and engineering group compiled design recommendations from FAQs related to AM processes and technologies. Video illustrations are included.

www.digitalengineering247.com/article/
design-guidelines-for-3d-printing-additive-manufacturing/

How to Design for Additive Manufacturing: Experts give their advice!

www.3dnatives.com/en/design-for-additive-manufacturing-expert-advice-121120195/

AM PROFESSIONAL ASSOCIATIONS, CONFERENCES, AND MEETINGS

AMUG – Additive Manufacturing User's Group
Dedicated to educating and advancing AM technologies.

www.amug.com/

AMUG Conference – Annual

www.amug.com/amug-conference/

RAPID + TCT Conference Event – Annual

www.rapid3devent.com/

National Additive Manufacturing Association (NAMA)

additivemfg.org/

Association for Metal Additive Manufacturing

my.mpif.org/MPIF/Associations/AMAM

Additive Manufacturing Association

www.vdma.org/en/v2viewer/-/v2article/render/15539539

Index

A

Abaqus ATOM/TOSCA, 54
abrasion-resistance, 72
ABS. *See* Acrylonitrile Butadiene Styrene
ACADEMI. *See* Advanced Curriculum
  in Additive Design, Engineering, and
  Manufacturing Innovation
accuracy. *See also* dimensional accuracy;
  standard accuracy; tolerances
  DED, 90
  geometry, finishing, 147
  vat photopolymerization, 100
Accurate Clear Epoxy Solid (ACES), 28
acrylate, 115
acrylics, 70–72
Acrylonitrile Butadiene Styrene (ABS), 69, 71, 80
acrylonitrile styrene acrylate (ASA), 70, 80
Additive Automation, 146
additive manufacturing (AM). *See also*
  three-dimensional printing; *specific topics*
  applications, 23–33
  benefits, 3–4, 22–23, 26, 32–33, 38, 160
  costs, 32, 142, 159
  hiring, 163–164
  history, 33–35
  implementation, 154
  job roles, 8–11
  manufacturing processes, 5–8
  process and technology, 85–103
  requirements, 153–165
  resources, 167–174
  technologies, processes, and materials, 11–13
  traditional manufacturing compared
    to, 9–11, 26
additive manufacturing design engineer, 15
Additive Manufacturing Format (AMF), 60–62
additive manufacturing process manager, 9
additive manufacturing specialist, 11
additive manufacturing technician, 10, 16–17
additive manufacturing technologist, 9

additive molding manufacturing engineer, 18
Additive Works Amphyon, 131
Adidas midsoles, 30–31
Advanced Curriculum in Additive Design,
  Engineering, and Manufacturing
  Innovation (ACADEMI), 162
aerospace industry, 26, 32, 67, 72, 73,
  90, 96, 98–99
aesthetic, 68, 76, 148
Agile Manufacturing Software
  Suite/3YOURMIND, 163
AI. *See* artificial intelligence
Altair Inspire, 54
Altair OptiStruct, 54
aluminum (alloys), 24–25, 72, 73, 76, 80
AM. *See* additive manufacturing
AmericaMakes, 34
AMF. *See* Additive Manufacturing Format
AMFG, 107, 163
amorphous thermoplastics, 79
AMUG, 162
anisotropic material, 46, 48, 55
ANSYS, 57, 131
application development technical service
  manager, 9
Arcam EBM Q20plus, 119
artificial intelligence (AI), 56
ASA. *See* acrylonitrile styrene acrylate
ASTM F42, 2
atomization, 72, 119
Authentise MES, 163
AutoDesk, 55, 131
automotive industry, 26, 32, 66

B

binder jetting (BJ), 7, 34, 68, 78, 88–90, 116
  ExOne, 29
  food, 94
  independent quality assurance, 126

Fundamentals of Additive Manfacturing for the Practitioner, First Edition. Sheku Kamara and
Kathy S. Faggiani © 2021 John Wiley & Sons, Inc. Published 2021 by John Wiley & Sons, Inc.